It has ever been my pleasure
To use mirth and truth together.

Grimmelshausen

Contents

5

Hans Jacob Christoffel von Grimmelshausen

13

The Baroque Novel and Grimmelshausen's "Simplicissimus"

23

Grimmelshausen's Life

35

Grimmelshausen's Work and His Subjects

51

Grimmelshausen's Sources

57

Grimmelshausen and His Publishers

65

Grimmelshausen's Influence

71

Views on Grimmelshausen

75

Chronological Table

91

Bibliography

Christoph Stoll

Hans Jacob Christoffel von Grimmelshausen
1676/1976

1976   Inter Nationes
Bonn-Bad Godesberg

About the Author

Christoph Stoll, born 1941 in Nuremberg, studied Germanistics and Philosophy in Erlangen and Munich, Associate of the Academy of Sciences and Literature in Mainz since 1969. Publications include a *Bibliographie der Personalbibliographien der deutschen Gegenwartsliteratur*, Munich 1971 (in collaboration with Herbert Wiesner and Irena Zivsa); *Sprachgesellschaften im Deutschland des 17. Jahrhunderts*, Munich 1973; *Ernst Kreuder. Von ihm — über ihn*, Mainz 1974 (in collaboration with Bernd Goldmann).

© 1976 by Heinz Moos Verlag, Munich
In collaboration with Inter Nationes, Bonn-Bad Godesberg
Translation: Patricia Crampton, London
Overall production: Kastner & Callwey, Munich
Printed in the Federal Republic of Germany

Hans Jacob Christoffel von Grimmelshausen

When, on 17th August 1676, the Pastor of the small Black Forest village of Renchen entered in his parish register: "Today there died in the Lord the estimable Johann Christoph von Grimmelshausen, a highly gifted, very cultivated man, Mayor of this place" — he certainly had no idea that this entry referred to by far the most important German novelist of the baroque period, one of the best known writers in all German literature. In other words, the author of the much read and frequently imitated novel "Simplicissimus, Melchior Sternfels von Fuchshaim", which appeared in 1668 under a pseudonym; it has survived the centuries as the only German-language baroque novel and continues to exercise the same power of attraction over readers throughout the world as it did 300 years ago. The "Simplicissimus" apparently owes its fascination to the fact that even when it was published it was bursting the boundaries drawn round contemporary German literature, and that by its realistic portrayal of its time it became a timeless work of art, standing well apart from its age, although a product of it. Grimmelshausen's "Simplicissimus" at once appears to the impartial reader as the practical visualization of an age which was overshadowed — long after its end — by one of the longest and creullest wars which Germany has ever experienced.

The Thirty Years' War, the causes of which lay far back in time and can ultimately be found in the religious schism initiated by the Reformation in Germany, began — with the "Prague Defenestration" of 23 May 1618 — as a religious war, a conflict between Catholicism and Protestantism, but increasingly expanded into a European power struggle on German soil. Germany became "the chess-board on which the great game was played out between the European opponents" (J. Haller). Sweden and France set the tone of the laborious negotiations which led in October 1648 to the conclusion of the longed-for Peace of Westphalia.

For three bloody decades mercenaries from every possible country fought under every conceivable standard. On the Protestant side it was mainly the English and the Scots, Danes and Swedes, Finns and Dutchmen who followed the drums of the recruiting squad. Spaniards, Italians, Walloons, Cossacks and Croats fought on the side of the Catholic imperial troops. The greatest vogue among the generals and mercenary leaders was enjoyed by Wallenstein, the Generalissimo of the Emperor Ferdinand II, who enforced a strict camp and battle discipline on his uncommitted mercenaries, but allowed them to plunder the areas they had conquered. In accordance with the slogan "War feeds war" the population of the occupied country had to bear all the costs of the war. Towns and villages were forced to pay completely prohibitive contributions if they wanted to be spared from looting. Monasteries and churches were pillaged and burned down. Marauders made the roads unsafe. Despite repeated decrees issued for the protection of the population, the most frightful excesses such as torture, looting, burning and rape became the common features of the war. So it was unavoidable that a more and more brutal soldiery, whose excesses became increasingly insufferable towards the end of the war waged against the Emperor by France and Sweden, was faced with a population which was often completely demoralized by sheer want following the disruptions of war. In addition to the devastation and destruction and to general famine there were the epidemics, above all the dreaded plague which raged through the armies as much as it did among the people of the towns and villages. At the end of that war — still present in the minds of Germans today, thanks to Grimmelshausen's "Simplicissimus" and Schiller's "Wallenstein" — Germany had lost half its population, great stretches of country were left waste and uninhabited, the national economy was ruined and morals had degenerated. From the experience of the war and its atrocities grew the pessimism characteristic of the era, an awareness of the vanity and transitory nature of this world. Fear of death, "memento mori" was the constant companion of the people of the baroque age, the hereafter was its determining factor. On the other hand, however, the awareness of their own nothingness and futility

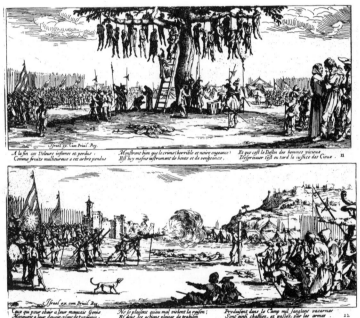

From the Cycle "Misères de la Guerre" by Jacques Callot (1633).

also gave birth to a burning hunger for life, a desire for the pleasures of this world, whose transience they recognized. People tried to master their passions and cupidity as well as the earthly adversities, their external insecurity and unsettled state, with "constantia", the principal virtue of Christian stoicism, the contemporary philosophy now grown popular after the rediscovery of stoical thinking in the Renaissance.

The conclusion of the Peace of Münster and Osnabrück, restoring the status of 1555, had temporarily put an end to the denominational wrangling, while it finally set the seal on denominational schism. Far from the denominational strife there had arisen at the beginning of the century, beside and against the controversial rigid Protestant orthodoxy, the Pan-Sophist outlook, nourished at the springs of medieval mysticism and Paracelsian natural philosophy, of the "philosophus teutonicus" Jacob Boehme, a cobbler from Görlitz, which was to exert an influence stretching far beyond the frontiers of Germany. In Boehme's enigmatically obscure work ("Aurora, that is: the rising dawn and mother of philosophies", 1612 et al), the alchemical, astrological and mystical religious knowledge of his period has been welded into an individual cosmological system. The unorthodox religiosity of the 17th century, ultimately leading to the broad movement of Pietism, was conceived in magical, alchemical origins and outlook. God and the world, the world and man, macrocosm and microcosm were brought into harmony.

In stark contrast to this world picture, with its strange mixture of natural knowledge and superstition, were the developments — and here too this century shows its charactcristic schisms, contrasts and upheaval — in the sphere of the natural sciences, largely influenced by the great English, French and Dutch philosophers, but still breaking ground with some hesitancy. Knowledge acquired long before was breaking through against considerable opposition. It was not until the middle of the 17th century that the revolutionary discovery by Nicolas Copernicus, the heliocentric world picture which he justified in his work "De revolutionibus orbium coelestium" of 1543, found general recognition — after Giordano Bruno had died at the stake in 1600 and after the work of Galileo

Galilei had fallen victim to the Inquisition in 1617. The new think-ing based on the natural sciences broke through primarily in astronomy — whose most important German representative was Johannes Kepler, who helped to justify the Copernican doctrine — mathematics (Descartes, Newton, Leibniz), physics (Pascal) and medicine, leading to countless important inventions and discoveries (the astronomical telescope, barometer, air pump, electrifying ma-chine, pendulum clock, the adding machine; discovery of the spec-trum and the human circulation). In epistemology and metaphysics, which was increasingly freeing itself from its role as the hand-maiden of theology, names such as Francis Bacon, Thomas Hobbes, René Descartes and finally Baruch Spinoza stand for the break with medieval scholasticism, which experienced a final flowering in the course of the Counter-reformation as "baroque scholasticism" under the influence of Spanish Jesuits at the German and Dutch Univer-sities, and for the rise of nationalism which was to become such a feature of the 18th century.

The most politically pregnant outcome of the Thirty Years' War was the downfall of the Empire and the established hegemony of France, whose absolutist form of government became the pattern for Prussia and Austria and the many small and minute territorial princedoms of Germany. With the expansion of absolutism and the political centralization peculiar to it, a far-reaching social reor-ganization was initiated. The estates in the traditional sense dis-appeared, farmers and the landed gentry were condemned to impotence by their impoverishment. The establishment of the mer-cantile economic system increasingly strengthened the bourgeoisie, which, with the farmers, had had most to suffer in the war. The courts, however tiny, increasingly developed into centres of new social and artistic life. The Prince, whether worldly or spiritual, enjoyed himself in the role of patron of a culture which was almost entirely supported by the bourgeois class.

Largely owing to the devastating war and political breakdown — there was no capital to act as an intellectual centre — there was never such a flourishing development of art and literature in Ger-many as there was in neighbouring France, at the court of the Sun king, which had become the glittering meeting point for the most important artists and writers of French classicism even under Cardinal Richelieu and then later, when Louis XIV came of age in 1651. Nevertheless, the fruits of the literary output in Germany in the 17th century — actually during the thirty years of the great war — should not be under-estimated. In particular, a great deal was taking place — if very belatedly in some spheres, as with the German literature of the time in general — in the sphere of linguistic theory, the consideration of the potential of the German language, which had to be emancipated in view of the predominance of European languages in Germany and what was regarded as "à la mode". Under the leadership of Prince Louis of Anhalt-Köthen, in 1617, a society of aristocrats and scholars was founded, calling itself the "Fruitful Society", copying the model of the Florentine "Accademia della Crusca" and concerned with the cultivation and purity of the German language. In 1624 Martin Opitz referred in his "Teutsche Poemata" and his "Buch von der deutschen Poeterey" to the indisputable equality of German. Then a whole series of grammars, poems, poetic anthologies, rule books and rhyme lexicons appeared as guides to scholarly familiarity with the Germans' own language. Though belatedly, and still in intimate dependance on Italian, Spanish, French, English and Dutch models, Germany was accordingly also able to develop a literature of quality in its national language, which became more and more courtly as the century hastened towards its end.

To today's observer the new German literature of the 17th century looks like a deliberately artificial literature, breaking with the tradition of popular literature which had been produced in considerable quantity by the age of the Reformation. The readers of this artifical poetry, these grandiloquent novels, the audiences of these dramas which were produced as major state operations, were to be found at the courts and among the high bourgeoisie. This was the literature of a cosmopolitan upper-class familiar with the classical tradition of antiquity and a modern education generally gained at foreign Universities. The courtly historical novel, which found its purest expression in Germany towards the end of the 17th century but had already outlived itself even then, was directly representative of this élitist literature.

Die Fruchtbringende Gesellschaft. Copperplate by Matthäus Merian from "Der Fruchtbringenden Gesellschaft Nahmen / Vorhaben / Gemählde und Wörter" (1646).

der Abenteuerliche
Simplicissimus Teutsch

Ich wurde durchs Fewer wie Phoenix geborn.
Ich flog durch die Lüffte! würd doch nit verlorn.
Ich wandert durchs Wasser, Ich raißt über Landt
in solchem Umbschwermen macht ich mir bekandt
was mich offt betrühet, und selten ergetzt,
was war das? Ich habs in diß Buche gesetzt,
damit sich der Leser gleich, wie ich izt thue,
entferne der Thorheit und lebe in Rhue.

Title engraving of the first edition of "Simplicissimus Teutsch" in 1668.

The Baroque Novel and Grimmelshausen's "Simplicissimus"

The 16th century had certainly brought in a rich harvest of prose novels abroad, but not in Germany. If we except Johann Fischarts' (c. 1546–1590) congenial version of François Rabelais' "Les horibles et espoventables faictz et prouesses du très renommé Pantagruel", published in 1575, which is still known today as a "gallimaufry", then Jörg Wickram (c. 1505 to before 1562) was the only one to have experimented successfully with the prose novel ("Der Goldfaden" 1554, etc.). His bourgeois novel had, however, already been displaced in the 16th century by the Amadis novel which was distributed through Germany from 1569 onwards in twenty-four weighty volumes. With the unparalleled success of this chivalric adventure novel, which conquered all Europe from its origins in Spain, a genre was born which was to remain characteristic of the representative part played by German novels in the baroque period: the courtly-historical romance. Thanks to the efforts of the reformer Martin Opitz (1597–1639) who had created the link between the German lyric and the artificial lyric from abroad, and thanks to the industrious translation activities of the "Fruitful Society", further works of this genre quickly found acceptance in Germany. Opitz himself translated in 1626 the "Argenis" of John Barclay, the content and form of which became the pattern for the German courtly romance which developed fully towards the end of the century; Philipp von Zesen — to name only one or two examples — translated "Die Afrikanische Sophonisbe" (1647) from a text by François du Soucy, Sieur de Cerzan; Johann Wilhelm von Stubenberg translated Madeleine de Scudery's "Clélie, historie romaine" under the title "Clelia: a Roman Story" (1664). Translations from the Italian rounded off the list of titles of foreign courtly historical romances, which are one and all written according to a precisely defined pattern derived from late Roman antiquity — possibly Meliodor's "Aetheopica". The centre of events in the novel is always occupied by two young lovers drawn from the highest ranks of the nobility who are separated by force, undergo countless setbacks during the separation, are refined by their trials and ultimately rewarded by marriage. The structure is more or less always the same as well: at the beginning of the novel the reader is taken straight into the middle of the story and learns the background only subsequently. The story-teller remains anonymous and has no contact with the reader.

The first of the courtly historical romances which was not a translation was by Andreas Heinrich Buchholtz (1607–1671). He gave it the typical title "The Miraculous Tale of the Christian German High Prince Hercules and the Bohemian Royal Maiden Valisca" (1659). Although Buchholtz's hereos still represented the knightly type à la "Amadis", yet the characters in the later and more complete novels of this genre were taken from the courtly society of the 17th century. This was the case in Heinrich Anselm von Zigler and Kliphausen's "Asiatic Banise" (1689), the most popular novel of the 17th century, and in the monumental artistic creations of Duke Anton Ulrich of Brunswick-Wolfenbüttel (1633–1714); "The most excellent Syrian Aramena", 1669–1673, "Octavia, a Roman Story" (1677–1707), but above all in Daniel Caspar von Lohenstein's three thousand-page work "Magnanimous General Arminius" which also appeared in 1689. This romance, in which the action itself is covered with countless digressions, lyrical and dramatic interpolations and thoughts on the natural sciences and morality — and therefore referred to by Eichendorff as an "encyclopaedia gone mad" — matchlessly fulfilled the demands made on the heroic novel.

The literary historians of the 19th century, lacking empathy and influenced by the aesthetics of the Goethe period — and Eichendorff speaks from this attitude — branded such creations as bombast, without recognizing the artistic ambitions of the baroque. It was not until the 20th century that the artistry of composition and language in these novels was appreciated once again.

The second of the prevailing types of baroque romance, the pastoral novel, also received considerable stimulus from foreign literature. And here the way for the "pastorals" in Germany was prepared above all by Honoré d'Urfé's "Astré", which appeared from 1607 to 1627. Once again Martin Opitz gave the direction, not this time with a translation but with his novel-like "Schäfferey von der Nimfen Hercinie", published in 1630, which was followed by a quantity of extremely popular works of this genre, such as Jakob Schwieger's "Die verführte Schäferin Cynthie durch Listiges Nachstellen des Floridans" (1660). In contrast to the supra-individual courtly historical romances, the pastorals always told a private love story, which — since love is not regarded here as something positive — almost always ends in the separation of the couple. The world portrayed is the uncourtly world of the small landed gentry.

In sharp contrast to both these types of romance, which were joined by various mixed forms and the type of the "gallant" and the "political" novel, is the picaresque novel, with which Grimmelshausen's "Simplicissimus" can be classified. Here too it is true to say that the "lower", picaresque novel is inconceivable in Germany without its foreign predecessors, some of them more than a century old. For instance, the bravura "Lazarillo de Tormes" by anonymous writer was published as early as 1554 — a socio-critical, satirical Spanish novel, the German translation of which (1617) sacrificed much of its originality to moralistic editing. The same thing happened to Mateo Aleman's "Guzman de Alfarache", published in 1599, "edited" by the counter-reformatory campaigner Aegidius Albertinus (c. 1560—1620) and published in German in 1615 under the title "Der Landstörzer: oder Gusman von Alfarache oder Picaro genannt". The Picaro Guzman, an out and out rogue, who stays on top in every possible situation — he is a priest's disciple, kitchen boy, soldier, goes into the service of a Cardinal, enters a monastery, which he flees in order to marry, becomes a pit boy in the Tyrol, is condemned to death and pardoned — this disturber of the peace becomes in the hands of Albertinus, librarian at the court of Maximilian of Bavaria, a remorseful and penitent sinner, and in this shape, a model for Grimmelshausen.

Besides the Spanish picaro — and Cervantes' "Don Quixote" is one — the German picaresque novel modelled itself on the French "roman comique", especially Charles Sorel's "Histoire comique de Francion" of 1623, which was translated twice in Germany: 1662 and 1663. The "low novel, in contrast to the courtly and pastoral romance, is set among the socially "irrelevant" classes, soldiers, actors, servants, beggars, robbers, whores. It gets its characters from the "outcasts" of the courtly society, on which it nevertheless abuts — generally at a critical and satirical distance. The intangible teller of the courtly tale is replaced in the picaresque novel by the first person story-teller, who makes his own experiences credible, describes these experiences in retrospect and communicates directly with the reader. Once again in contrast to the courtly type, the structure of the novel is linear, the life-story of the hero being recounted section by section, from the beginning, although there is no lack of digression here too. Apart from this, the picaresque novel always relates to the present and reflects the social conditions of its time of origin.

It was long after the translations of the "Lazarillo", the "Guzman" and the "Picara Justina" had become familiar that Germany made its connection with the genre of the picaresque novel — and then it was with an unsurpassed masterpiece, which was both a modification and an enlargement of the Spanish novels: Grimmelshausen's "Der abenteuerliche Simplicissimus Teutsch". One year before Anton Ulrich of Brunswick's "Aramena", twenty years before Zigler's "Asiatic Banise" and Lohenstein's "Arminius" appeared, the "Simplicissimus", as the most popular literary work of the 17th century, defined the precise counter-position; it brought about the "rejection of the ideal of elegance" (P. Bockmann) of the baroque era and thus initiated the development of the realistic bourgeois novel in Germany.

The hero of the title describes in retrospect his lively career, which begins and ends in Spessart. As a nameless boy, he is brought up by simple Spessart peasants, driven from the farm of page of the German translation of y's "Arcadia", Frankfurt/Main 1629.

ARCADIA
Der Gräffin von Pembrock.
Das ist:
Ein sehr anmütige
Historische Beschreibung
Arcadischer Gedicht vnd Geschichten/
mit eingemängten Schäffereyen vnd
Poesien.
Warinn nicht allein von den wahren Eygen-
schafften keuscher vnnd beständiger Liebe gehandelt/ sondern
auch ein lebendig Bild deß gantzen menschlichen Wesens vnd
Wandels/ auffs zierlichst für Augen ge-
stellet wird:
Allen Hoff- Raths- Kriegs- vnd Weltleuten/ Edel vnd Vn-
edel/ Hohes vnd Niderstands Personen/ die hin vnd wider/ sonder-
lich aber an Herrn Höfen/ handeln vnd wandeln/
lieblich/ nützlich vnd nötig zulesen:
Anfangs in Englischer Sprach beschrieben/ durch den weyland Wolge-
bornen/ Trefflich-beredten vnd Berümbten Englischen
Graffen vnd Ritter
H. PHILIPPS SIDNEY:
Nachmalen von vnterschiedlichen vornehmen Personen ins Frantzösi-
sche; Nun aber auß beyden in vnser Hochdeutsche Sprach/
fleissig vnd trewlich übersetzt
Durch
VALENTINVM THEOCRITVM von Hirschberg.
Mit schönen newen Kupfferstücken gezieret.
Gedruckt zu Franckfurt am Mayn/ bey Caspar Rötell/
In Verlegung Matthä Merian.
ANNO M. DC. XXIX.
Mit Röm. Kays. Mayt. Freyheit auff 6. Jahr.

Der Landtstörtzer:
Gusman von Al-
farche oder Picaro genannt/
dessen wunderbarliches/ abenthew-
rlichs vnd possirlichs Leben/ was gestallt er
hier alle ort der Welt durchloffen/ allerhand Ständt / Dienst vnd Aembter ver-
richt/vil guts vnd böses begangen vnd auß-
gestanden/ jetzt reich / bald arm/ vnd wider-
umb reich vnd gar elendig worden/ doch
letztlichen sich bekehrt hat/hierin
beschriben wird.

Durch
EGIDIVM ALBERTINVM,
Fürstl: Durchl: in Bayrn Secretarium,
theils auß dem Spanischen verteutscht/
theils gemehrt vnd ge-
bessert.

Getruckt zu München / durch Ni-
colaum Henricum.

ANNO M. DC. XV.

page of the translation of "Gusman Alfarache" by Ägidius Albertinus, ch 1615.

EL INGENIOSO
HIDALGO DON QVI-
XOTE DE LA MANCHA,
Compuesto por Miguel de Ceruantes
Saauedra.

DIRIGIDO AL DVQVE DE BEIAR,
Marques de Gibraleon, Conde de Benalcaçar, y Baña-
res, Vizconde de la Puebla de Alcozer, Señor de
las villas de Capilla, Curiel, y
Burguillos.

Año, 1605.

CON PRIVILEGIO,
EN MADRID, Por Iuan de la Cuesta.

Vendese en casa de Francisco de Robles, librero del Rey nro señor.

page of the first edition of "Don ote de La Mancha" by Cervantes, rid 1605.

his supposed parents by looting soldiers and taken up by a hermit who gives him the name Simplicius because of his limitless simplicity and instructs him in the Christian faith and in reading and writing. After his death he goes as page to Governor Ramsay, who has the simple youth changed into a calf and makes him play the fool. During the chaos of the last two stages of the Thirty Years' War, whose horrors he experiences with intensity, the childishly naive pure fool becomes the noted-notorious "Huntsman of Soest", a warrior prey to every pleasure, carried up and down by his adventurous destiny. At one time on the peak of fortune, the once pious, now profligate hero falls into the deepest misery after an episode in the Paris "Venusberg": now he has to spend his life as mountebank and musketeer, even as one of a marauding gang, before settling, after a pilgrimage, as a peasant in the Black Forest, where he finally discovers his noble origins. A fresh start takes him to Moscow, from where he sets out round the world. After his return he lives as a hermit, renouncing the world.

Grimmelshausen's masterpiece, told with a humorous and satirical objectivity, is a splendid example of that tension, typical of the baroque age, between the now and the hereafter, between addiction to the world and rejection of the world, between sin and penitence; it shows the impermanence and transistoriness of that world, by portraying an individual destiny in all its heights and depths in rich colours. Apparently loosely linked together from numerous single episodes and several fine allegories, the novel, comprising five volumes in the first version, actually has a well-considered structure, diverging from the prevalent baroque novel, which has stimulated researches into making very controversial interpretations. At an early stage parallels were seen in the structure of the "Simplicissimus" to Wolfram von Eschenbach's (1170—1220) medieval verse novel "Parsival" (1200—1210), whose hero, like Simplicius, grows up as a pure and oppressed nameless fool and steps out into an alien world in a fool's costume. However, some parallels in motif simply admit of the reserved observation "that the 'Parsival' can hardly have had any influence beyond the first volume of the 'Simplicissimus'" (G. Weydt). There were numerous, if generally unsuccessful attempts to interpret "Simplicissimus" as an educational or evolutionary novel, on a par with Goethe's "Wilhelm Meister" and Gottfried Keller's "Grüner Heinrich". Recently people have been trying to interpret Grimmelshausen's novel and the principles of its composition by means of structures borrowed from the baroque itself. G. Weydt, for instance, was able to give convincing indications of the astrological structure of "Simplicissimus": the novel runs through the seven phases of the planets Saturn, Mars, the Sun, Jupiter, Venus, Mercury and the Moon, which in turn guide the destiny of the hero.

The many and often highly contradictory interpretations of Grimmelshausen's "Simplicissimus" in the past and today make it clear that it — like all the great creations of literature — must be regarded as an inexhaustible work of art from which there will constantly be new insights to be gained. Its liveliness and plasticity are due above all to the rare, happy accident that the author largely draws the adventurous life-story of his hero from his own experience, with the vision and knowledge of detail of someone who has actually "been there".

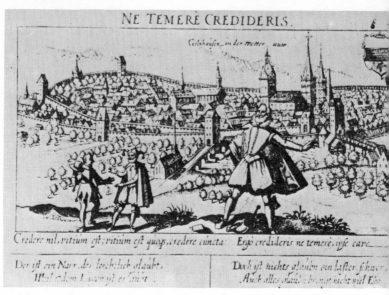

Gelnhausen. Copperplate by D. Meiner from: "Poetisches Schatzkästlein" (1623).

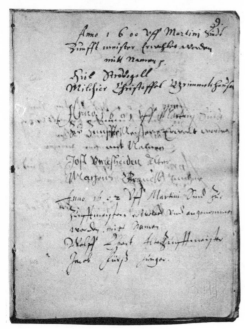

Entry in the Gelnhausen "Bäk-
ker- und Müllerzunftbuch"
(Guild Register of Bakers and
Millers) of 1595: The poet's
grandfather became Master of
the Guild in 1600.

20

Gelnhausen, Grimmelshausen's birthplace.

The Imperial Palace in Gelnhausen.

The house where Grimmelshausen was born in the Schmidtgasse, now the Hotel "Zum Weissen Ochsen".

Grimmelshausen Memorial in the Gelnhausen Municipal Gardens.

Grimmelshausen's Life

"Simplicissimus" and other works by the author offer countless points of reference for the biography of the young Grimmelshausen. In particular G. Könnecke has researched these with scrupulous care and for the period of the Thirty Years' War, they can be regarded as the sole sources for Grimmelshausen's life story. For as long as the identity of the author of "Der abenteuerliche Simplicissimus Teutsch" was unknown, the biography of the hero Simplicius was also thought to be that of his creator. The novel was published in 1668 under the pseudonym "German Schleifheim von Sulsfort", giving an invented place of publication and an equally invented publisher's name. It was Ernst T. Echtermeyer who finally, in 1838, unmasked the ten pseudonyms used by Grimmelshausen as anagrams of his name. This left the way open to further research into the life of this Hans Jacob Christoffel von Grimmelshausen, which still contains many question-marks today.

Whereas the still nameless hero of the novel was born in a secluded farm in Spessart shortly after the battle at Höchst, from which the commander of the League, Tilly, emerged the victor on 22 June 1622, Grimmelshausen probably came into the world in March 1621, in the free town of Gelnhausen, with its 1500 inhabitants, near Frankfurt am Main, between Spessart and Vogelsberg. A profound influence on his childhood was that of his grandfather Melchior Christoph, descended from a noble family in Thuringia, who took care of the little boy after his father's premature death and his mother's re-marriage. He had a house in the Schmidtgasse where he ran a wine tavern. From 1627 to 1634 Grimmelshausen seems to have attended the Lutheran Latin School in Gelnhausen. The battle

23

of Nördlingen on 8th September 1634, in which once again the hero of one of Grimmelshausen's novels, Springinsfeld, fought on the side of the victorious Imperial troops, brought about a decisive change in the boy's career. The Swedish Commander, Duke Bernhard of Weimar, was defeated in this battle by an Imperial and Bavarian superior force. Troops of Ferdinand, the General Infanta of Spain, poured to the north after the victory. On 13 September 1634 Gelnhausen was occupied and plundered by Croats from this army. In 1685 the Town Council was still recounting in emotional words "that this town was plundered in 1634 by the army of the Cardinal Infanta in their withdrawal to the Spanish Netherlands, the citizens being in general slaughtered, the rest driven away and the town itself set on fire and so devastated that the same remained for a considerable time uninhabited in view of the further mortality which followed" (this was the plague). The young Grimmelshausen must have fled with the majority of the population from his home town to the nearby fortress of Hanau.

At the beginning of 1635 he was probably picked up by a Hessian patrol and taken to Cassel and in July 1635 he was probably with the troops of the Cassel-Hessian Commander-in-Chief, General Holzapfel, called Melander. Grimmelshausen experienced the second capture of the "cloth and straw town with earthen walls" which had been destroyed by Tilly in 1631, just as on 4. 10. 1636 he was an eye-witness of the Battle of Wittstock. His description of this encounter in "Simplicissimus", which is among the masterpieces of Grimmelshausen's narrative art, according to H. Geulen is not in fact to be regarded as his completely original work "but as the conversion of a corresponding battle scence from Sidney's 'Arcadia'" (G. Weydt): "Our Profos did in fact stay quite a long way behind the battle with his people and the prisoners; at the same time, however, we were so close to our brigade that we could recognize everyone from behind by his clothes; and when a Swedish squadron encountered our own, both we and those fighting were in deadly peril, for in a moment the air was so filled with balls singing over our heads that it looked as if the salvoes were being fired in our honour, the fearful ducked down as if they would have liked to hide inside themselves; but those who had courage

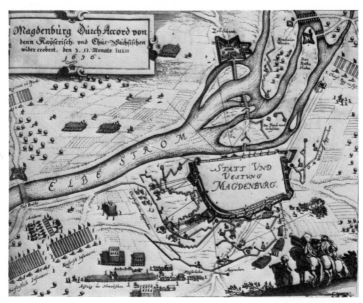

The Reconquest of Magdeburg by Imperial and Saxon Electoral troops on 5 July 1636. Anonymous Copperplate from "Relationes Historicae Continuatio" (1636).

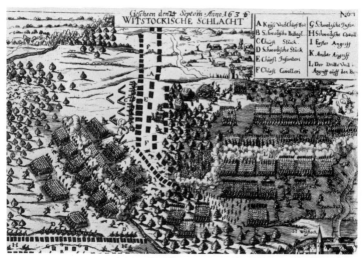

The Battle of Wittstock on 4 October 1636. Copperplate by. N. Weishun from "Abbildung der denckwürdigsten Schlachten" (1666).

and had been present at similar jests, let them pass over their heads without turning pale. But in the encounter itself each one sought to avoid his death by slaying the first one who attacked him, the horrible shooting, the rattle of harness, the cracking of the pikes and the cries both of the wounded and of the attackers, combined with the trumpets, drums and pipes, made a terrible music." So runs a part of Grimmelshausen's graphic description of the battle of Wittstock — certainly a paradigm for many battles without local colour during the Thirty Years' War.

There is a big difference when we come to the description of Westphalia. Grimmelshausen's detailed knowledge of Westphalian specialities, his familiarity with Westphalian landscapes, places and buildings, enable us to conclude with reasonable certainty that he stayed in Westphalia in 1637, probably as a groom in the regiment of Life Dragoons of the Imperial Field-Marshal Hans Count of Götz, who was stationed in Soest with his troops from the end of 1636 to March 1637 and with whose Master General of Ordnance, the Bavarian J. C. von der Wahl, Grimmelshausen probably undertook extended expeditions through Westphalia to Hesse and Waldeck.

The next reference point for the author's biography is the Upper Rhine fortress of Breisach. With Count Götz's regiment, who had been ordered by the Bavarian Elector Maximilian to relieve the fortress which was under heavy siege by Bernhard von Weimar, Grimmelshausen may have come to South Germany. After a siege of more than four months, Breisach was forced to surrender on 17. 12. 1638. Count Götz was unable to intervene in Breisach, since his regiment had suffered a crushing defeat on 9 August 1638 at Wittenweier. The General was cited in an electoral patent of 8 November in Munich for alleged incompetence, dismissed on 3 December and imprisoned in Ingolstadt. He was not rehabilitated until August 1640. Meanwhile his troops went into winter quarters in Swabia. According to the meticulous researches of G. Könnecke, it is almost certain that Grimmelshausen took part in Count Götz's campaign against Bernhard von Weimar as a soldier in the regiment of Life Dragoons. The conclusive evidence for this fact is supplied by two first-person stories from the "Everlasting Calendar", one of

Left: Johann Wenzel Count Götz. Copperplate by Joh. Chr. Sysang from: "Conterfet Kupferstich" (1722).

Right: Joachim Christian Count Wahl. Copperplate by Petrus de Jode after Anselmus van Hulle (1658).

Soest. View from the Southwest. Copperplate by Matthäus Merian from: "Topographia Westfaliae" (1647).

which begins as follows: "When in my 17th year I was still a musketeer or dragoon and after the summer had passed and the campaign was ended, I was in winter quarters in the country of those peoples who still wear a codpiece in the manner of the ancient Teutons as a sign of their inherent constancy . . ." The first sentence of the other anecdote, entitled "Der Teutsche Bauer" runs: "I was once ordered to Swabia with a party of Götz's army, which at that time lay in Neustadt in the Black Forest". The knowledge of the Swabian custom of wearing codpieces and the fact that Count Götz had pitched camp in Neustadt in the Black Forest from 26 September to 23 October 1638 — of which neither the novel nor any source used by Grimmelshausen reported — indicates that he was an eye-witness and that the "I" of the stories must be indentical with himself.

In 1639 we find Grimmelshausen in the newly-acquired regiment of Freiherr Hans Reinhard von Schauenburg, Commandant of the heavily embattled free city of Offenburg from 1638 — a man whose influence was to be of great importance to the further career of the writer. Once again it is one of the little stories from the "Everlasting Calendar" which seems to confirm Grimmelshausen's presence in Offenburg at that time:

"After defeating Breysach, Duke Bernhardt von Weimar prepared to besiege Offenburg as well, where the Imperial Colonel von Schauenburg was in command. At that time a turbot was caught in the millstream, which was held to be an extraordinary miracle and was presented by the fisherman to the said Colonel, who ate it. But a very young musketeer, coming from Gelnhausen, made this prediction: 'The town of Offenburg', he said, 'would not be taken as long as the Colonel lived and commanded it.' For which the youth was greatly mocked. In fact it was found that he had spoken the truth, since the Colonel held the town until peace was concluded. Accordingly such things are not all to be despised."

Promoted by the Commandant, Grimmelshausen then became assistant clerk in his regimental staff office as assistant to the Secretary, Magister Johann Witsch. Once again he had to leave the countryside which was to be of decisive importance to him and his work, the Black Forest and the Upper Rhine. With H. R. von

Left: Hans Reinhard von Schauenburg. Portrait in oils (1648).

Right: Rent book kept by Grimmelshausen while he was employed as steward by the Schauenburg family.

Schauenburg's brother-in-law, Colonel Johann Burkhard Elter, he moved to Bavaria in 1648, now apparently as Regimental Secretary. In Wasserburg on the Inn, Elter was in the service of the Bavarian Elector Commandant. "It may have been Elter who, as a vassal of the Elector of Bavaria, acquainted the author with the writings of the Munich Court Librarian Aegidius Albertinus and his translations" (G. Weydt).

Meanwhile, the Peace of Münster and Osnabrück had been concluded and Grimmelshausen was released from service as Regimental Secretary. After his return to Offenburg in 1649 he married on 30 August the twenty-one year-old Catharina Henninger, daughter of a cavalry sergeant from Saverne and in the same year

29

The Swedish Peace Banquet in Nuremberg Town Hall from Johann Klaj "Irene" (1650).

became Steward to his former regimental Commander Hans Reinhard and his cousin Carl Bernhard von Schauenburg in Gaisbach, where he lived in Hans Reinhard's Steward's house. Gaisbach is now part of the town of Oberkirch in Ortenau, which in Grimmelshausen's time belonged to the territory of the Bishop of Strasbourg. The Steward's duties were to supervise his master's properties, to collect the dues and taxes of the farmers dependant on the Schauenburgs from Gaisbach and the rest of the Renchtal, to conduct judicial and extra-judicial proceedings and to undertake many official journeys on his behalf. During the eleven years of his Stewardship he had to draw up a detailed annual account every year, of which a few in Grimmelshausen's own hand have been preserved.

In the course of the years the Steward became relatively prosperous. The acquisition of several properties in Gaisbach and the building of two houses on one of these properties, the "Spitalbühne", are evidence of this. In 1656 he leased the Steward's house of Freiherr Philipp Hannibal of Schauenburg which was more practical for his work and opened a wine tavern in it, which he called "At the Silver Star".

By 1658 Grimmelshausen was landlord of the "Silver Star", as his second calling; three years later he left his position with the Schauenburgs, but in 1662 he had taken another post quite close by. He became Castellan and Administrator at the Ullenburg, which the fashionable Strasbourg doctor Johannes Küffer had received in 1661 in lieu of fees. After four years in the service of this wealthy doctor Grimmelshausen moved back in 1665 to his Gaisbach inn, in order to occupy himself with the office of Mayor from that place. The publication of his first works came in the Gaisbach years. In 1666 he published his "Satirical Pilgrim" a kind of tract, in which the antithetical 'answerability' of all life's questions is dealt with according to the dialectical method. The same year saw the publication of the heroic-gallant novel "Virtuous Joseph", the first German 'Joseph' novel, a version of the stories from Genesis.

On 9 February 1667 the author applied to the Strasbourg Bishop Franz Egon Count Fürstenberg for the position of Mayor in Renchen, which had fallen free and — after his father-in-law Johann Henninger had vouched for him — he obtained it in April, although he could not take it up until June. From then until his death he was virtually a servant of the State, whose duties included exercise of the lower police powers and jurisdiction, keeping the property registers and collecting taxes.

Louis XIV of France's wars of conquest against Holland and Germany from 1672 to 1678 once again involved the border area around Strasbourg in the turmoil of war. Once again there were marching armies, billeting and compulsory dues. As recently discovered documents show, Mayor Grimmelshausen intervened vigorously in the interests of his district, with the Strasbourg Bishop among others. The immediate area of Renchen became a theatre of war in June and July 1675 and on 23 and 24 July the place itself was occupied by the French Commander Turenne. Once again Grimmelshausen, shortly before his death, had to do military service, not as a professional soldier but as the leader of a kind of citizens' army. The entry in the Renchen Parish Register is not clear in this respect: "On 17 August 1676 there died in the Lord the estimable Johann Christoff von Grimmelshausen, a highly gifted, very cultivated man, Mayor of this place. Although he had volunteered for military service owing to the disturbances of war and his children were scattered in all directions, nevertheless all of them gathered here owing to a dispensation. And their father died, fortified by the pious reception of the sacrament of the Eucharist and was buried. May his soul rest in Holy peace."

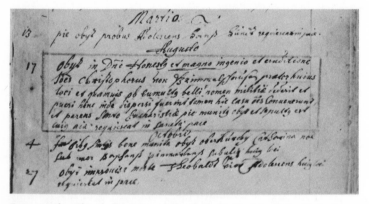

Entry in the Church Register of the Catholic Parish of the Holy Cross, in Renchen on the death of Grimmelshausen on 17 August 1676.

Record of Grimmelshausen's Marriage on 30 August 1649 in the Church Register of the Parish of the Holy Cross in Offenburg.

As we find in the marriage register and many signatures from the author under official documents, so this entry by Pastor Caspar Meyer in the Death Register of the Catholic Parish of Renchen shows that Grimmelshausen had called to mind the prefix of nobility which had no doubt been dropped by his bourgeois grandfather and from a certain point in time had once again appended it to his name. We can assume that this happened about the time of his marriage. As we also see from the Death Register, Grimmelshausen, born in Protestant Gelnhausen, died a Catholic. The same thing applies to his conversion as to large sections of his biography: when and where it happened cannot be stated with any certainty.

Der Seltzame Springinsfeld. Title plate from the first collected edition of 1683/84.

Grimmelshausen's Work
and His Subjects

When, one must ask in connection with this far from typical biography of an author, did the much-occupied regimental secretary, administrator, landlord and mayor find the time to write more than twenty books, each and every one of which appeared within a period of less than ten years? The fact that he found this possible "is among the mysteries of this unique life" (G. Weydt).

Apart from the "Simplicissimus", which was followed in 1669 by a Robinson Crusoe pastiche as a sequel, the works of 1666 already mentioned and the Simplicianic writings, which are about to be discussed, in 1670 the romantic idyll "Dietwald and Amelinde" was published under the author's name, dedicated to Philipp Hannibal von Schauenburg — the complicated love story of two royal children in the fifth/sixth century. These were followed by the anti-Macchiavellistic tract "Ratio Status" and in 1672 by the novel of a self-abnegating love: "Proximus and Lympida", both works under Grimmelshausen's own name. Apart from this, the writer was a fruitful composer of calendars ("The European Wonder Tale Calendar" 1669, "The Everlasting Calendar" 1671).

But down the centuries all that have remained are the "Simplicianic writings" of which Grimmelshausen himself regarded the "Courasche", the "Springinsfeld" and the "Magic Bird's Nest" more or less as sequels to his major work. Like the smaller "Sim-

Bertolt Brecht: "Mother Courage and her Children. A Chronicle of the Thirty Years War." Premiere 1941 in Zürich under L. Lindtberg with Therese Giehse in the lead role.

plicianic writings" ("Der erste Beernhäuter" and "Gauckel-Tasche" 1670, "Verkehrte Welt", "Rathstübel Plutonis" 1672, "Teutscher Michel" 1672, etc.) they are more or less loosely linked with "Simplicissimus" through their characters or subjects.

"More mobilis than nobilis", Simplex called the lady who met him in the mineral springs at Griesbach and who was alleged to have been the mother of his son. In order to revenge herself for Simplex's defamatory story, she dictated to one "Philarchus Grossus von Trommenheim", the dissolute story of her life. Libuschka — which was the real name of Courasche — was brought up by a foster-mother near Budweis. The illegitimate daughter of a Bohemian Count, dressed as a boy, became involved in the tumult of war, becoming the lover or wife of a series of cavalry and infantry captains, but also of a simple musketeer, and worked for some time in the sutler's trade with that Springinsfeld, who is also well known from "Simplicissimus". Finally, as the wife of a lieutenant of gypsies, she traversed the whole of Europe.

Title plate of the first edition of "Courasche" of 1670.

In the story within a story of the "Seltzamer Springinsfeld", the deceit of Courasche is revealed to a company assembled in the inn and it is proved that — in spite of Courasche — Simplex is the father of the son foisted on him by her. Present during this interchange are an old man with a beard: Simplex, a peg-leg: Springinsfeld, a vintner and a steward (here Grimmelshausen brings himself into the story) as well as the foster parents of Simplex. During the night Springinsfeld recounts his life story: as a "fine, diverting juggler's boy", moving about with his father through the fairs of Europe, he too became involved in the war, through which he served as a drummer, musketeer and pikeman, but also as a rich dragoon in many battles, including the battle of Nördlingen. He met with a lieutenant's widow, Courasche herself, with whom he made common cause. He fought in distant theatres of war against the Turks, lost a leg in Crete and returned to Germany. Meanwhile he had met and married a lyre player, who had come into possession of a bird's nest which made its possessor invisible. The "lyrist"

Left: Title plate from the first edition of "Rath-Stübel Plutonis" (1672).
Right: Das wunderbarliche Vogel-Nest. Titleplate to the first edition of part I (1672).

abused her new powers and was burned. Springinsfeld wandered through Germany as a cripple with his fiddle.

"The Magic Bird's Nest" of the lyre player fell into the hands of a soldier when she was burned. He tells what happened to him as the owner of the nest with the power of invisibility: a mass of observations of people of many classes who thought themselves unobserved. In his hands the bird's nest prevents injustice and makes intervention possible where help is needed.

In the second part of the story a merchant secures the nest and in contrast to his intelligent predecessor, uses it principally for sexual escapades. Despite his ability to make himself invisible, he is severely wounded as a soldier in the war beetween the Netherlands and France. Brought to understanding by the influence of a Catholic priest, he renounces his magic powers and throws the bird's nest into the Rhine between Strasbourg and Kehl.

As this modest summary of the content of the stories in the cycle of the "Simplicianic Writings" makes obvious, one of Grimmelshausen's central themes is war. The act of war is, at it were, the foil against which his great narrative gifts unfold. War, for the writer of Simplicissimus, has nothing to do with heroism, with fame and glory. "The splendid deeds of heroism would be worthy of high praise had they not been achieved with the death and injury of other people." He describes war just as he himself had experienced it, all its terrors and cruelties, from the viewpoint of the simple soldier, carried now up, now down by the vagaries of war, of which the famous allegory of the tree of the states says:

> *Hunger and thirst, and heat and cold,*
> *Labour and poverty, so we're told,*
> *Acts of injustice, violation,*
> *These are a trooper's occupation.*

This rhyme is all the less of a lie because it assorted well with the facts, for eating and drinking, suffering hunger and thirst, whoring and sodomy, rattling and gambling, carousing and revelling, murdering and being murdered, killing and being killed, plaguing and being drilled again, chasing and being chased again, frightening and being frightened again, robbing and being robbed again, plundering and being plundered again, fearing and being made to fear, causing misery and suffering miserably again, beating and being beaten again; and in sum, only spoiling and injuring and in return being spoiled and injured again, was their whole doing and being; in which they were not prevented by winter or summer, by snow or ice, by heat or cold, by rain or wind, by mountain or valley, by field or swamp, by ditches, passes, seas, moors, water, fire, or ramparts, by father or mother, sister or brother, danger to their own bodies, souls and consciences, yes, even by loss of life, proceeded industriously in their actions until in the end, bit by bit, in battles, sieges, assaults, campaigns and even in quarters (though they are the soldiers' earthly paradise, especially when they meet fat farmers), they fell, died, rotted and perished; but for just a few, who in their old age, if they had not greatly oppressed and stolen, made the best beggars and adventurers of all."

The wild doings of the soldiery, which Simplex sees breaking into the solid world of Spessart and himself experiences as a forager with the Croats, as the huntsman of Soest and as a member of a marauding gang himself, stands in strong contrast to the profoundly religious nature of the introductory and concluding chapters, which are dominated by ideas of withdrawal from the world and asceticism. In the last chapter of the "Continuatio" of 1669, which finally brings Simplex as a hermit to a distant island, a Dutch captain recounts the motives which led the hermit to refuse to sail back with the Dutchmen: "As long as he had been with people in the world, he had always received more vexation from enemies than from friends and even friends often gave rise to more inconvenience than friendship; if he had no friends here to love and serve him, equally he had no enemies who hated him, both kinds of people being capable of bringing every man to sin, but dispensing with both he could the better serve God in peace; at first he had certainly both suffered and withstood many temptations both from himself and from the arch-enemy of all men, yet at all times through God's grace he had found help, consolation and salvation in the wounds of his Saviour (who was his only refuge)."

The religious attitude demonstrated in Grimmelshausen's works seems very distant from the denominational schism of his time. In the central Jupiter episode of Book 3 this wise fool explains "how he will unite the religions with each other": "after my hero has made universal peace for the whole world he will deliver a very moving sermon to the spiritual and worldly leaders and heads of the Christian peoples and various churches and make strikingly clear to them the previous highly injurious controversies in matters of faith and by means of highly reasonable arguments he will bring them of themselves to desire a common unity . . ." This hero would speak to a huge ecumenical council of the most learned and pious theologians in the world, summoned by him, and "explain that, as quickly as possible and yet with the most mature and industrious deliberations, they must firstly put aside the disputes which exist between their religions and subsequently, with true unanimity must set down in writing the right, true, holy and Christian religion in

Simplicius Simplicissimus, Book I, Chapter 19. Etching by Erich Erler (1921).

Der wahn
betreugt

Der Thor sucht Trost in Eitelkeit.
Der Klug in Gott die himlischfreud.

Simplicius Simplicissimus, Book I, Chapter 6. Etching from the "Baroque Simplicissimus" of 1671.

accordance with Holy Scripture, the ancient tradition and the acknowledged Holy Fathers." In a dispute with the reformed vicar of Soest the huntsman acknowledges: "You know quite well, vicar, that I am a Christian ... but apart from that I confess that I am of neither Peter nor Paul (i.e. neither Catholic nor Lutheran), but believe with the utmost simplicity what the twelve articles of the general divine Christian belief contain ..." Thus in relation to a hopeless war, whose ultimate causes lay in denominational hostility, Simplex reveals himself as the representative of an irenic, simple Christianity, though still oriented towards Catholicism. The song put in the mouth of the hermit — i.e. the father of Simplex — is an impressive proof of such simple, uninhibited piety. The first two verses run:

> *Come, boon of night, O nightingale,*
> *And let your voice's joyful scale*
> *Its loveliest cadence ring.*
> *Come! On your Maker praises heap,*
> *Since other birds have gone to sleep*
> *And not again will sing:*
> > *Better than all*
> > *Can your voice in loud rejoicing*
> > *Tell of love*
> *Four our God in heaven above.*

> *Although the sunshine now has gone*
> *And we in darkness must live on,*
> *Still we can always sing*
> *Of all God's goodness and His might,*
> *No power to stay us has the night,*
> *Still shall His praises ring.*
> > *So above all*
> > *Let your voice in loud rejoicing*
> > *Tell of love*
> *For our God in heaven above.*

Despite the renunciation of the world which concludes "Simplicissimus", the Simplicianic writings are absolutely teeming with exuberant life. The often crude and sometimes gross, but always satirically exaggerated realism of the description of the everyday life of soldiers and citizens shows the author to be a man who had both feet firmly on the ground. His undisguised sympathy, in the literal sense of the word, went to the "invalid" people, the artisans, day wage earners, people on the fringes of society and the peasants, who gave to the social whole impressively realized in the allegory of the Tree of States "its strength, and were always ready to give

Simplicius Simplicisimus, Book I, Chapter 15. Line drawing by Josef Hegenbarth (1970).

again whenever it was lost; in fact they replaced the deficit of the fallen leaves from their own, to their own still greater ruin". The still nameless "little boy" learned the following song in Spessart:

> *You, much despised peasant band,*
> *Are yet the finest in the land.*
> *A man need only look at you*
> *To give the praise which is your due.*
>
> *How would it be now in the world*
> *If Adam had not tilled the field.*
> *Over his furrows once did bend*
> *The man from whom all kings descend.*
>
> *Almost all things are under you,*
> *For everything the earth can do*
> *To feed and nourish all the land*
> *Passes before then through your hand.*
>
> *The Emperor, he whom God gives*
> *To guard us, in the first place lives*
> *Through your hands, and the soldier too,*
> *Though great the harm he does to you.*

This has nothing to do with the glorification of the peasant life; the song really expresses the author's partisanship, very unusual in the seventeenth century, for a class which he saw realistically and which at that time really was "much despised", oppressed and exploited. The majority of contemporary writers knew the peasants only as characters in their idyllic pastorals or in the "shepherd or rural plays, presenting peasant life, and called satyrical". So said the "Poetischer Trichter" of Georg Philipp Harsdörffer. The steward of the Schauenburgs, who knew the "vine men" of the Renchtal in this capacity and later on as an innkeeper and mayor, was in the best position to understand their situation, which was particularly critical after the war. Such direct knowledge of the conditions can be clearly detected from an episode in the first part of the "Bird's Nest": the invisible owner of the nest is there when an emissary with two soldiers seizes the last possession of a peasant, his goat. "The woman beat her hands over her head and screamed on and on without interruption: 'Oh may God in heaven have mercy! How am I to feed my poor little children now?' ... But the man begged the executor before God and for God, for the sake of his passion, for the sake of all his Saints and of the Last Judgment for but eight days' grace; but in vain." The father of eight children, robbed of his goat, describes his situation: "I have not a cent of money only to buy the dear salt! I have no fat only for a water soup! For all my bitter toil and work there is scarcely enough, thanks to the unbearable oppression, to feed my poor children only with the dear dry bread! And after all this, these extortioners come and take the best from my house! What am I to do now, if you are to be robbed of both, the butter and the milk?"

However, to try to simplify Grimmelshausen because of his support for the under-privileged of his time into a kind of spokesman for the peasants and tribune of the people, is out of the question, even if only because of the author's efforts, pointed out by M. Koschlig, to present himself in the aristocratic circles in which he also moved as the "noble Herr von Grimmelshausen", whose Simpliciana, in these very circles, apparently did not find the recognition he had hoped for. So when he still published "the old manuscripts of 'Ratio Status', 'Dietwald and Amelinde' and 'Proximus and Lympida', it was certainly not for their literary content but — as his full, undisguised name on these three particular writings demonstrates — for the sake of his personal reputation, which he painfully missed and therefore so passionately desired" (M. Koschlig).

At the same time, it is true that the author's heart belonged to his popular "Simplicianic Writings". And just as he knew of the needs and troubles of the "invalid people" so he also spoke and wrote their language. "He had 'looked the people in the mouth' and made use of his sterling resources of style, which give his writings that vividness and plasticity for which they are still read today. The first editions of the "Simplicianic Writings" reveal him as absolute master of numerous German dialects, as fluent in Alemannic as he is in Hessian and Westphalian. In the "Teutscher Michel" he makes no secret of his familiarity with the various German linguistic areas: "In the little places towards Prague as good a German is spoken as anywhere in all Germany; this means that the German speakers have no peasant neighbours in the surrounding villages who ruin their speech for them, whereas, on the contrary, the Frankfurters learn from the Wetterauers, the Strasbourgers from the Kocherbergers, the Tübingers from the Swabians, the Regensburgers from the Bavarians, the Marburgers from the Hessians, the Leipzigers from the Meissners and also many others from their Low German speaking neighbours, many a blemish, although they include many quite learned people, even academies full of young students, who all learn an elegant German; since the people have more to do with the peasants than with the scholars. But of all the major German towns I think none speaks a more senseless German than the otherwise majestic Cologne, whose speech is otherwise better suited to no one than the womenfolk, but only to those who are also beautiful." It was the later editors who — no doubt with the best intentions — modernized and standardized his writing.

The standard had been set in the 17th century by the grammars and orthographies of the influential "Fruitful Society" — an association of noblemen, scholars and writers who stood principally for the purification of the German language from the countless foreign words which had penetrated it, owing in particular to the war. It was founded on the model of the Italian "Accademia della Crusca" in 1617 and adopted its ceremonial and customs. On admission every member of the society was given his own society name and a symbol from the plant world. Grimmelshausen could not resist mockery; in the "Everlasting Calendar" he wrote: "Italian society. Among the Swiss he saw a variety of donkeys and mules coming over the mountains from Italy with lemons, oranges and all kinds of goods; so he said to the brother of his heart, look, for God's sake, that is the fruitful society of the Italians." Despite this occasional mockery, he was basically in sympathy with the purist tendencies of the linguistic societies, only he was critical of the immoderate exaggerations of such a one as the young Philipp von Zesen, who was a member of the "Fruitful Society" from 1648 onwards, who wanted to ban loan words from the German language, saying "daylights" for windows, for instance, and "maidenholder" for convent. The author was also very critical on other occasions of this Zesen, who had become his rival as the writer of the Joseph novel "Assenat" (1670). In "The Bird's Nest", for instance, he positively tears Zesen's work to pieces.

The publication of the book "The Pomp and Swagger of the World-Famous Simplicissimus with his German Michel", in which Grimmelshausen deals with problems of the German language and the efforts of the linguistic societies and their "language heroes", aroused confusion, especially in the societies themselves. On January 1673 Quirin Moscherosch, a member of the Nuremberg "Shepherd and Flower Order on the Pegnitz", an author with the publisher Felssecker and a parson in the neighbourhood of Strasbourg, wrote to Sigmund von Birken: "The well-known Simplicissimus, erstwhile my neighbour, and only a humble local magistrate, but a Dauss Ess (a ruffian) and homo Satyricus in folio, has had a

Deß Weltberuffenen
SIMPLICISSIMI
Pralerey und Gepräng
mit seinem
Teutschen
Michel/
Jedermänniglichen / wanns seyn
kan/ ohne Lachen zu lesen erlaubt
Von
Signeur Meßmahl.

Title page from the first edition of "Teutscher Michel" of 1673.

Title plate of the first edition of "Teutscher Michel".

little tract published by. H. Felssecker before Christmas, entitled The Linguistic Pomp of German Michel, in the style of the painter's mixture of colours, in which he most satirically mocks the German language heroes; I would dearly like to know when someone is going to stop his mouth with a couple of sheets; if my office allowed it I would not fail to do so. But I believe that he has stirred up a wasps' nest which will soon revenge itself on him . . ." The wasps' nest stayed quiet, perhaps partly because in these slightly élitist circles they did not take much notice of a man such as Grimmelshausen, who was after all concerned with the peasants and their language.

However, it has long been demonstrated that — although he stood so far apart from these associations — Grimmelshausen owed a great deal to the work of some of their prominent members. But this raises the question of the poet's sources.

PIAZZA VNIVERSALE,

das ist:

Allgemeiner Schauwplatz/ oder Marckt / vnd Zusammenkunfft aller Professionen/ Künsten / Geschäfften / Händlen vnd Handtwercken/ so in der ganzen Welt geübt werden:

Deßgleichen

Wann/vnd von wem sie erfunden: Auch welcher massen dieselbige von Tag zu Tag zugenommen:

Sampt außführlicher Beschreibung alles dessen/ so dar, zu gehörig: Beneben der darin vorfallenden Mängel Ver, besserung/ vnd kurtze Annotation zur jeden Dis, curs insonderheit.

Nicht allein allen Politicis, sondern auch jedermänniglich wes Standts sie seynd/sehr lustig zu lesen.

Erstlich durch

Thomam Garzonum auß allerhand Authoribus vnd experimentis Italiänisch zusammen getragen / vnd wegen seiner sonderlichen Anmühtigkeit zum offtermal in selbiger Sprach außgangen.

Nunmehro aber gemeinem Vatterlandt Teutscher Nation zu gut auffs trewlichste in vnsere Muttersprach vbersetzt/

Vnd so wol mit nohtwendigen Marginalien, als vnterschiedlichen Registern geziert.

Gedruckt zu Franckfurt am Mayn / bey Nicolao Hoffman / in Verlegung Lucä Iennis.

M. DC. XIX.

DILIGENTIA

FLEIS BRINGT NAHRVNG

EXPERIENTIA

ZEIT BRINGT ERFAHRVNG

Title page of the first German translation of T. Garzoni's "Piazza Universale", Frankfurt 1619.

Grimmelshausen's Sources

The extent of Grimmelshausen's culture — the village priest of Renchen had described him as "highly gifted and very educated" — was and is one of the controversial themes of Grimmelshausen research. Although some of them regarded him as highly educated, that is, the master of many foreign languages and extremely well read in the literature of the 16th and 17th centuries, to others he seemed to be a superficially educated "peasant poet" or "village philosopher", educated only in popular literature. A passage from the "Satirical Pilgrim", frequently interpreted as a personal statement by Grimmelshausen, suggested the second view: "As we know, he himself had studied, learned and experienced nothing, but as soon as he had barely grasped the ABC, went into the wars, becoming a snotty musketeer at the age of ten, growing up lumpishly any-old-where in this same ludicrous life, without good discipline and instruction, like any other coarse rogue, ignorant donkey, a dunce and an idiot . . ." The man who wrote this was far from being an ignorant donkey. In fact he had gained a solid foundation of basic knowledge at the Gelnhausen Latin School, whose requirements were very high, continued consistently with his education and made himself extremely familiar with the European literature of his time — if in part only through literary intermediaries. The works demonstrably or presumably used by him would amount to a library of not less than 170 books, some of them in many volumes. We can hardly assume that Grimmelshausen possessed such a library — even in view of his not exactly rosy economic siuation. On the other hand, the libraries of the Schauenburgs — for whom he worked for a long time after his resignation, as the editor of Klaus von Schauenburg's "German Peace Council" (1670) — offered him abundant reading matter, as did the book collections of the cosmopolitan Dr. Küffer at the Ullenburg, who was not unfamiliar with the literary scene — not to speak of the close vicinity of Strasbourg, an important centre for printers, publishers and book dealers.

Historical dates, facts and places from the period of the Thirty Years' War, which are mentioned and made use of in "Simplicissimus" and still more in "Courasche" and "Springinsfeld", could be found by Grimmelshausen in the "Theatrum Europaeum", abundantly adorned with etchings by Matthäus Merian. This was a kind of universal chronicle, the third volume of which had the pretentious sub-title: "Historical Chronicle Part 3, containing a short and truthful description of all the events and stories worthy of consideration and chronicling, which have taken place throughout the world, in both the East and West Indies, and especially in Europe, in France, Spain, Italy, Great Britain, Denmark, Sweden, Poland, Bohemia, Hungary, Transylvania, Wallachia, Moldavia, also parts of Turkey and Barbary, etc. In High and Low Germany, but most of all in the Imperial German nation, etc., in the last six years, from anno 1633 to 1638 inclusive, in churches, world government and warfare, on every hand ..." Another historical source — among many others — was the "Revised German Florus" of Eberhard von Wassenberg, which appeared in 1647. Among the numerous contemporary German literary figures whose influence on Grimmelshausen's work is unmistakeable, the thorough research of G. Weydt and his school has found that the most important was Georg Philipp Harsdörffer, a Nuremberg patrician, man of letters, member of the "Fruitful Society" and the Hamburg "Germanist Society", co-founder and chairman of the "Pegnitz Flower Order", as the author of the proverbial "Poetic (Nuremberg) Road to Literature", art expert of his age and connoisseur of the whole European literary scene, master compiler and intermediary of the literary themes and subjects prevailing in the belles-lettres and historiography of the first half of the century. His widely distributed "Conversation Pieces for Women" and others of his voluminous collections of "selected" stories, maxims, etc., such as the "Pegnitz Shepherd Poem" were the storehouses from which Grimmelshausen drew materials and subjects for many of the central episodes of "Simplicissimus" and also for his Calendars. He also drew on the "Parsival" — often named in one breath with his own major work — by Wolfram von Eschenbach, which he knew from an edition of 1477, on the Luther Bible, the German popular books (e. g. "Eulenspiegel", "Fortunatus", "Faust" and "Melusine"), on the humorous literature of the 15th and 16th centuries and on the poems of the Nuremberg cobbler poet Hans Sachs. The most fruitful influence on the little "Simplicianic Writings" was Johann Michael Moscherosch, born in Willstädt near Strasbourg and literarily far superior to his already well-known brother Quirin, with his "Stories of Philander von Sittewald".

Only a few names and titles will be mentioned here of the fictional library available to Grimmelshausen, above all the Picaresque novel "Vida del pícaro Guzmán de Alfarache" by Mateo Alémán, known to Grimmelshausen in the translation by Ägidius Albertinus of 1615, to which "Simplicissimus" owed at least the "autobiographical form of the novel, the technique of arranging the adventures and episodes and thereby the possibility of a continuation" (V. Meid). A French representative of the coarsely realistic "Roman comique", once again Charles Sorel's "La Vraie Histoire Comique de Francion", directed against the pastoral novel, written in 1623, translated into German in 1662 and 1668, was an important influence on Grimmelshausen's narrative style. The apologia for the secluded rural life from the pen of the Spanish Bishop Antonio de Guevara, the "Menosprecio de corte y alabanza de aldea", published in 1539 and translated into German by Albertinus in 1604, was also taken up by Grimmelshausen, with many other authors. The "Adieu, World" passages of the fifth book of "Simplicissimus" were taken directly from the German translation. Whether and in what edition the writer knew Cervantes' "Don Quixote" is uncertain — despite demonstrable subject borrowings, which of course might also have been obtained via Harsdörffer. The same applies to the "Picara Justina" produced in 1605 by the Spaniard Andreas Perez, translated into German in 1620 under the

Saturnus. Woodcut by Georg Pencz from the series "Die sieben Planeten und ihr Einfluss auf die Menschenkinder" (1531).

title "Landstörtzerin Justina Dietzin" which was long regarded as the model for "Courasche".

What is quite certain since J. H. Scholte's detailed investigations is the use throughout the whole Simplicianic work of the "Piazza Universale" of the Italian Thomas Garzoni, the contents of which are set out as follows in the German translation of 1619: "General show-place, or market and meeting-point of all professions, arts, businesses, dealings and handicrafts practised throughout the world . . ." So this book is an encyclopaedia; Grimmelshausen owed it a great deal of his inspiration. Some also came from English literature, for instance Henry Neville's "Isle of Pines", translated into German in 1668, an early Robinsonade, from which Grimmelshausen took themes and motifs for several chapters of his "Continuatio" which appeared a year later. Of the "serious" travel literature of his time, he knew the repeatedly published eye-witness report by Adam Olearius "Journey to Moscow", later "Revised Description of the Muscovite and Persian Journey" which he used in the fifth book of "Simplicissimus" as the basis for his hero's adventurous world trip.

Besides belles-lettres, encyclopaedias and works of travel literature, as Koschlig has convincingly shown, Grimmelshausen used straightforward textbooks. For instance, the "Oeconomia Ruralis et Domestica" of M. Johann Colerus, which was an accurate introduction to good economy in house, farm and field. Finally, he must also have been familiar with theological, alchemical and above all astrological works; in the astrological field, particularly with books containing pictorial illustrations, on which Weydt comments: "Not previously proved to be one of Grimmelshausen's sources, but owing to its wide distribution very likely to have been one, at least representative of a refined type of the descriptions of planetary gods known to him, are the etchings of Dürer's pupil Georg Pencz from the first half of the 16th century". Weydt explains the astrological structure of "Simplicissimus" on the basis of these etchings.

This brief glance at Grimmelshausen's fictional library, which was much more extensive, makes it clear that the "ignorant donkey, dunce and idiot" was extremely well-read and adapted what he learned from his reading in masterly fashion when integrating it with his own work. It is only in very rare cases that he makes a simple take-over. Instead Grimmelshausen took the themes, motifs and materials of his sources — as the example of the description of the battle of Wittstock from Sidney's "Arcadia" clearly illustrates — reorganized them with complete originality and integrated them with the mainstream of his tale with genius. In this way, the author of the single German baroque novel which has actually been read all through the centuries, also became the transmitter of a chapter of European literary history which is otherwise little known in Germany.

Title plate of the first edition of "Proximus und Lympida" (1672).

Grimmelshausen and His Publishers

This is the title of the book by M. Koschlig, which is important to the extremely confused Grimmelshausen philology. Koschlig has made it his life's task to clarify the relationship of the author with his publishers and has been able to bring astonishing results to light in quite recent years. With Grimmelshausen as his example, he has been able to shed light on publishing practice and accordingly on a piece of the fundamental literary life of the baroque period. The controversy over legitimate editions of "Simplicissimus" in particular, or of the ones unauthorised by the author, is not yet over, so here, as in the writer's biography, there are a lot of question-marks.

In his very first publications, the "Satirical Pilgrim" and the "Virtuous Joseph", Grimmelshausen had extremely bad luck. Apparently he had long reached an agreement with the Strasbourg publisher Johann Christoph Nagel on the publication of these works by his house, when publishing rights and accordingly his actual living were withdrawn from Nagel, partly through the intrigues of his fellow-publishers and guild members. So Grimmelshausen had to look for a new publisher and he found one for the two parts of the "Satirical Pilgrim", probably by the good offices of the Nuremberg calendar and newspaper publisher Wolf Eberhard Felssecker, in the famous Leipzig publisher Georg Heinrich Frommann, who published them in 1666 and 1667 respectively and again in 1671. Leipzig, which was then as now, with Frankfurt, the centre of the book trade and an important meeting-place for publish-

Ratio Status. Title plate from the first edition of 1670.

ers, must have led to the contacts between Frommann and the still unknown Nuremberg publisher. The "Virtuous Joseph" appeared in Autumn 1666, published by Frommann's Nuremberg partner Felssecker, who could not at that time have suspected that with the author of this courtly love-story, which was little in demand, he had acquired himself — as we would say today — a best-selling author. The rise of his publishing house is indivisibly linked with the name of Grimmelshausen. It began with the extraordinary success of the first edition of "Der abenteuerliche Simplicissimus Teutsch", dated 1669 but actually published a year earlier. From then on "Simplicius" and "Simplicianic" were to become more or less literary trade-marks in which the ambitious Felssecker and his heirs scented big business and achieved it — often against the will and without the knowledge of the author. Felssecker (to be found under the pseudonym "Johann Fillion"), following up the huge demand, pushed out a second edition in Autumn 1668 or Spring 1669 which was obviously prepared in great haste and is therefore very slovenly. More than that — with this second edition there appeared, both separately and in the same binding, the "Continuatio", that is the sixth book of the "Simplicissimus".

Felssecker's overwhelming success with this book, which came onto the market under a pseudonym, with obviously fictitious names for place and publisher and without even the protection of a simple licence, was converted into cash profits for himself by a Frankfurt publisher who was always close to bankcruptcy and who also had an old debt to pay off against Felssecker, owing to a Nuremberg pirated edition. Georg Müller, the Frankfurter, tried to save himself from imminent ruin with a pirated edition of the book, which he announced in Autumn 1669 under the following title: "Simplicius and Simplicissimus newly arranged and corrected, or the description of an extraordinary vagabond, where and how he came into the world and what he saw, learned and did there". Corrected — at least in the modern view — was no description for this "Simplicissimus", although it had indeed been "newly arranged". While the first edition of the novel was the work of an author who used dialect — either deliberately or unconsciously — as a stylistic medium, Müller's newly-arranged version was quite different. The impoverished publisher had set to on his own and at breakneck speed to standardize and modernize Grimmelshausen's racily exuberant language, which to his mind was not at all easily understood. Piquantly enough, Müller had recourse to the standardizing works of a man called Christian Gueintz, written under the auspices of the "Fruitful Society", by which Grimmelshausen did not set much store. Known — very characteristically — as "the Orderer" in his society, the works in question were entitled: "Project for German Language Teaching" (1641) and "German Orthography" (1645) — a grammar and an orthography.

Müller's edition, very attractively entitled by J. H. Scholte the "Schoolmaster-Simplicissimus", had an astonishing effect. After Felssecker had re-published his own "Simplicissimus" — again in great haste — in Autumn 1669, with a good selling title, in order to squeeze Müller's pirated edition out of the market, simply dropping all the magnificently announced innovations and expansions mentioned in the title, as another stroke against Müller (who had reprinted and "corrected" the "Courasche" in 1671 as well — he published something like a luxury edition, this time giving his own name as publisher, in Autumn 1671. In a foreword put into the mouth of Simplex, this edition is claimed to have been "caused by a bold and very audacious imitator, in that he had seen fit to snatch this little work of mine, put only into the hands of my publisher, snatching from him his highly praiseworthy care and expense, industry and labour, on the first edition and agreeable presentation of my work, and had quite unlawfully appropriated it for himself". This "baroque Simplicissimus" (another apt expression of Scholte's) was provided "with twenty attractive copper-plates and three continuations" and its linguistic form was — that of Georg Müller's pirated edition, in other words, standardized along the schoolmasterly lines of Christian Gueintz and still further edited in addition.

With regard to this "luxury edition" the question arises of the illustrations to Grimmelshausen's work, which in this respect were at first handled in highly niggardly fashion. Besides some most impressive title copper-plates — above all the enigmatic chimera of the Simplicissimus edition, a composite of man, devil, fish and bird — the first editions had no other illustrations, if we exclude the "Beernhäuter" and in particular the "Gaukeltasche" which make use of some card pictures from Jost Amman's card game book of 1588, which were already almost a hundred years old and mutilated besides, and a woodcut taken and wrongly copied from Sebastian

Three etchings from the "Baroque Simplicissimus" of 1671: Book II, Chapter 7; Book IV, Chapter 22; Book V, Chapter 1.

Brant's much older "Ship of Fools" (1494). It was only the competition by the Frankfurt pirated edition which seems to have stimulated W. E. Felssecker to give his "baroque Simplicissimus" expensive copper-plates. They were not an unqualified masterpiece, but by comparison with the truly "baroque" illustrations of the three posthumous collected editions, they are of an almost lovable simplicity, very appropriate to the "Simplicius", such as, for instance, the portrayal of the friends on a pilgrimage towards recluseship. Eighteen of the twenty copper-plates carry the caption "Folly deceives" and one passage in the "Bird's Nest" made it seem probable at first "that the poet was intensely involved with the illustration of his work and that undoubtedly the conception of the pictures and possibly even part of their execution stemmed from him" (J. H. Scholte). The question as to whether the "baroque Simplicissimus" actually reached the public with Grimmelshausen's consent, as a kind of "last-hand edition" is — once again — disputed; according to M. Koschlig's most recent discoveries, however, as good as answered.

Now a new figure appears, whom one must assume to have been highly profitable to Grimmelshausen and his literary heritage as regards the publisher Felssecker: this was Johann Christoph Beer, a Lutheran theologian, who had not been accepted as a preacher and therefore became a reader for Felssecker and his sons. He produced an imposing miniature edition of G. P. Harsdörffer, with a considerable body of publications of his own and others. A superb compiler, a man with the right nose for big business, in short, a shrewd man of letters. Thus he was the exact counterpart to Grimmelshausen, who more or less covertly opposed Beer, but resigned himself in the end and — changed publishers. We can quite safely ascribe to Beer, 1. the publication and editing of the "Baroque Simplicissimus", the sixth edition of the novel which was certainly printed behind the author's back in 1672 and the three collected editions which appeared from 1683 to 1713, with a commentary by him. 2. The authorship of the "European Wonderful History Calendar" of 1670–1675, long ascribed to Grimmelshausen. Beer's autocratic actions, undoubtedly with the consent of the publisher, must have caused Grimmelshausen to break with Felssecker and to put his new books into another publisher's hands. This was the house of Georg Andreas Dollhopf in Strasbourg, in other words quite close to Grimmelshausen. Whether the "Rathstübel Plutonis" of 1672 — a conversation piece between Simplicianic and non-Simplicianic characters in the Bad Griesbach mineral spring, where they discussed the art of getting rich — and the "Teutsche Michel", already well known, were published in Strasbourg has not yet been proved, since there is no indication of any publisher. On the other hand, it is certain that both parts of the last great Simplicianic work, "The Magic Bird's Nest", the smaller works such as "Stoltzer Melcher", "Bart-Krieg" and "Galgen-Männlin" and the idyllic novel "Proximus and Lympida" were published by Dollhopf. Two of these works — "Bird's Nest I" and "Der Stoltzer Melcher" did also appear in Nuremberg — apparently pirated by Grimmelshausen's old publisher Felssecker, who had the effrontery to sue his Strasbourg rival at the Town Council as the alleged illegal publisher of these works, on 16 December 1672.

All this is sufficient to show that even in Grimmelshausen's lifetime publishers were squabbling over his books, because the reading public was fighting for them, in particular for the major work "Der abenteuerliche Simplicissimus Teutsch", which — after the three collected editions of 1683/84, 1685/99 and 1713, probably edited by J. C. Beer — appeared on the market from 1756 onwards in countless, again generally "edited", and frequently illustrated editions and still enthral the reader of today.

Wolff Eberhard Felssecker, Grimmelshausen's Nuremberg Publisher.
Copperplate by J. A. Böner with verses by Grimmelshausen.

Grimmelshausen's Influence

Beer was not to be the only who immediately recognized the favourable market position for the literary brand name "Simplicianic". In the wake of Grimmelshausen's extraordinary success with the public a large number of books appeared between 1670 and 1744 with this brand name in the title, in a bid for custom. Borrowing from the familiar expression "Robinsonade", these have become known to research as "Simpliciades". Most of them — satires, stories of adventure, collections of tales, etc. — displayed no connection with Grimmelshausen apart from the well-known brand name. Three novels of varying quality constitute an exception: "Dess Frantzösischen Kriegs-Simplicissimi Hochverwunderlicher Lebenslauff" ("The Wonderful Life of Simplicissimus in the French Wars") which appeared anonymously in 1682/3, which was joined in 1683 by the "Hungarian or Dacian Simplicissimus" by the Göppingen teacher and musician Daniel Speer. Here the author refers to his "two cousins, the German and French Simplicissimus". The well-packed novel of the composer, musician and actor Johann Beer, only later rediscovered as author, "The Simplicianic World Observer", appeared as early as 1677/79 and his more important "Jucundus Jucundissimus" quite deliberately imitated Grimmelshausen's title.

The contemporary readership of "Simplicius Simplicissimus" was apparently drawn from every possible social class. Prominent figures such as Duchess Sophie of Hanover and the universal genius Gottfried Wilhelm Leibniz were no less enchanted with the work than the public at large. After the appearance of the last of the three collected editions Grimmelshausen's major work was published another four times in the 18th century. Among its readers were Gotthold Ephraim Lessing, who spoke of the writings of "Samuel Greifenson".

Despite the total success of the "Simplicissimus", the novel and the other "Simplicianic Writings" were not really rediscovered until the turn of the 18th and 19th centuries by the German Romantics, whose catholicizing tendencies seized upon a character such as that of the hermit with joy. His fervent song to the nightingale attracted their admiration above all. Since its inclusion in the first edition of Achim von Arnim's and Clemens Brentano's famous collection of folk and art songs "Des Knaben Wunderhorn", an anthology of German lyrics without the song of the hermit is inconceivable. The Romantics were equally fascinated by such out of the way subjects and characters as the magic mandrake root from the "Galgen-Männlin", the "Bärenhäuter" (Lazybones) and others, which were taken up by Brentano, de la Motte Fouqué, Tieck and Eichendorff, and also by the brothers Jacob and Wilhelm Grimm in their "Children and Home Fairy-Tales" and the "German Legends".

The real rediscoverer of Grimmelshausen, the writer Hermann Kurz — who should not be confused with the early publisher of a scholarly edition of the "Simplicianic Writings", Heinrich Kurz — wrote a "Simplicissimus" of his own in 1836 and the Weinsberg senior physician and spiritualist Justinus Kerner wrote a "Lazybones in the Salt Spring" in 1835/37. Then came the first attempts to produce textually sound editions. Gustav Freytag made particular use of the "Courasche" and the "Springinsfeld" as historical sources for his much-read "Pictures from the German Past" which appeared in 1859/64. Towards the end of the century Richard Wagner's son Siegfried made an attempt at a "Lazybones", and a far lesser known Walter Freiherr von Rummel at a drama, "Simplicissimus", in 1901.

For almost half a century, a high quality newspaper of political

Grimmelshausen Memorial in Renchen (1879).

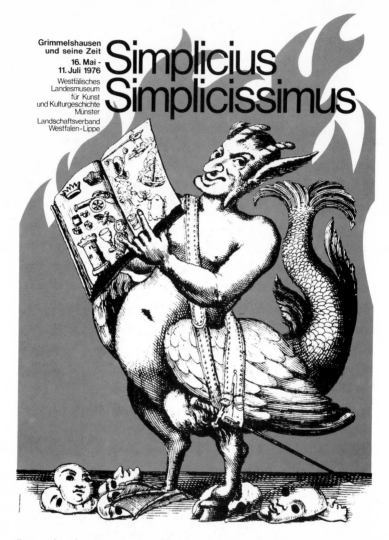

Grimmelshausen und seine Zeit

16. Mai - 11. Juli 1976

Westfälisches Landesmuseum für Kunst und Kulturgeschichte Münster

Landschaftsverband Westfalen-Lippe

Simplicius Simplicissimus

Poster for the great commemorative exhibition "Simplicius Simplicissimus. Grimmelshausen and his time". The exhibition is being shown in Münster, Westphalia from 16 May to 11 July 1976; in Oberkirch from 31 July to 24 August 1976; in Gelnhausen from 7 August to 4 September 1976; in Wolfenbüttel from 28 August to 10 October 1976; in Offenburg from 24 September to 3 October 1976; in Nuremberg from 7 November to 2 January 1977; in Karlsruhe from 21 January to 28 February 1977; in Soest from 20 March to 1 May 1977.

satire recalled Grimmelshausen's work in its title: the "Simplicissimus" which appeared from 1896 until it was suppressed and extinguished by National Socialism in Munich. The comparativist and expressionist poet Ernst Stadler, who was killed at the beginning of the First World War, published in "Der Aufbruch", his poetry collection of 1914, an unorthodox poem with the title "Simplicius Becomes a Hermit in the Black Forest and Writes his Life Story". Twenty years later the première of the chamber opera "Simplicius Simplicissimus" by the Munich composer Karl Amadeus Hartmann was performed. During the National Socialist era Grimmelshausen's work — even and especially by the Germanist School — was trimmed to "blood and soil". "It was possible to announce as a new research conclusion that Grimmelshausen's philosophy was rooted in the peasantry — but only by painfully suppressing or trivializing Grimmelshausen's many anti-peasant statements ... The fourth chapter in the third book of 'Simplicissimus' exercised an undeniable power of attraction, 'of the German hero, who will master the entire world and create peace between all nations'. German hopes of world domination found their confirmation here — in the mouth of a lunatic". (E. Lunding, 1950). The popular Black Forest bard Hermann Eris Busse, who became known at that time for his major work "Peasant Aristocracy" and was promoted by the National Socialists, adopted the writer with a monograph and the "Grimmelshausen Tale" "The Silver Star". It appeared in 1940 when emigrant circles had already long remembered Grimmelshausen. In the exile newspaper "International Literature — German Pages", which for a time called itself the "Central Organ of the International Association of Revolutionary Writers" and appeared from 1931 to 1945 in Moscow, Grimmelshausen also appeared with German classicists and Opitz under the heading "Documents of the Past". The texts selected under this heading "brought to light the enlightened progressive line of tradition of German history and literature and could largely be read as a deliberate allusion to the situation in the Third Reich" (H. A. Walter).

During his New York and London exile, Bertolt Brecht turned the character of "Courasche" into the heroine of one of his most-performed plays: "Mother Courage and her Children. A chronicle of the Thirty Years' War", which had its première in Zurich on 19 April 1941 with Therese Giese in the principal role. In his "Novel of a Novel. The Origins of Dr. Faustus", published in 1949, Thomas Mann reported that he had read Grimmelshausen's "Simplicissimus" as "companion reading" when he was in exile in America, during the writing of his Dr. Faustus in 1944. Traces of this reading can be found throughout the novel. In the same year, 1944, he wrote a foreword to the first Swedish translation of the "Simplicissimus", which he praised as "a narrative of magnificence".

This — and it was late enough — paved the way for Grimmelshausen into world literature. Between 1950 and 1960 translations of "Simplicissimus" appeared in Finnish, French, Japanese, Czech, Serbo-Croat, Italian, Flemish and Polish; a Russian translation followed in 1967.

Grimmelshausen's great success in Germany in the second half of the 17th century was repeated at international level in the middle of the 20th. This writer, completely unknown to his contemporaries, though they knew his "Simplicius Simplicissimus" very well indeed and valued it highly — Grimmelshausen, in fact, has thus become, beyond the bounds of the German language area as well, what he had always been, from 1668 onwards: an author with a place in world literature.

Grimmelshausen's signature

Views on Grimmelshausen

A book in the German language has been recommended to me, entitled Sinplisis Sinplisissimos. It was printed in Mompelgard; the title leads one to suspect the prince of that place as its author. I believe it is of equal rank with Bertoldo and Bertoldine; I therefore recommend it to your lady wife on condition that she takes the trouble to send me a copy; I believe that it is quite certainly to be found in Frankfurt.

Duchess Sophie of Hanover to her brother Karl Ludwig of the Palatinate, 5 November 1670

At the same time I have received the letter you have done me the honour to send me through the post and the one which arrived together with Madame Courage. This good lady was never so splendidly seated as on the casks of an excellent wine and although this gift — as also the lemons and oranges — was really for my lord husband, I could not prevent myself from sticking my nose and tongue into it in his absence and assembling small morsels from this area ... As his marriage partner, I proffer you my most respectful thanks in his name and mine, as also for the pretty adventure of Madame Courage: she knew how to trick the cavaliers. The story of Sinplicissimo begins very piously; I do not know if it ends in the same way; I shall have it read aloud to me while I prepare garters à la mode for my lord husband.

Duchess Sophie of Hanover to her brother Karl Ludwig of the Palatinate, 10 December 1670

On the second day of Easter the custom is, so I am told, that the priest tells a little tale which is called the *Oster-mährle* (Easter story). This time the Jesuits had taken one from a German book written to be laughed at and called Simplicissimus, and which is quite close to the genius of Francion; but it had been changed a little.

Gottfried Wilhelm von Leibniz to Duchess Sophie of Hanover, April 1688

We had spoken of Simplicissimus and you were of the opinion that it should be left to Easter 1802; since then, from idleness, I have not looked for another publisher and am now inclined to leave the whole thing; for those who can understand it it is there, the others will only be annoyed by it.

Ludwig Tieck to Friedrich Frommann, 1801

One of the most excellent books is the Simplicius Simplicissimus, Tieck has lent it to me, Lord, it's divine! It is rare, so in the next few days I will put an advertisement for it in the National Advertiser and will give a commission for this to be answered to Marburg, because I may not be here any longer then, and you will then buy it for me, as offered, and as there are many other books associated with it, then you will buy every single one which belongs to it; there is a lot about it in Koch.

Clemens Brentano to Fr. K. Savigny, 17. 6. 1803

... the undoubtedly very remarkable Simplicissimus, one of the most-read books in the second half of the 17th century, has nothing to do with scholarly education but is created from a humble view of customs and contemporary history.

August Wilhelm Schlegel, 1803

A reading from Simplicissimus. Goethe said of it that it was cleverer and more attractive in construction than Gil Blas. Only they, publisher and public, could not find an end.

Johann Wolfgang von Goethe in conversation, 1809

In Grimmelshausen we become acquainted ... with a writer who clearly recognizes both tendencies of his age, without confusing them with one another; a writer, gifted with a rare wealth of wit and imagination who commands the still rarer power and flexibility of mind to pursue the most varied literary tendencies in all their originality, so that he adds to them a not inconsiderable intellectual content from his own literary powers; who in both these tendencies knows how to handle the right form of presentation, suited to the content, with equal mastery, though poorly applied on one side; who undoubtedly stands alone and unequalled in the popular literature of his time and yet again shares all the defects and weaknesses of a predominant bad taste.

W. A. Passow, 1843

To be the publisher of the first Swedish translation of the adventures of Simplicissimus is a privilege and an honour on which I would like to congratulate you sincerely. It is quite surprising that this work has never yet been offered to the Swedish public in their own language, since it is not only an incomparable cultural document of the German baroque but is also full of memories of Sweden's own era of military heroism. It is a monument of the rarest kind to literature and to life, which has survived almost three centuries in all its freshness and will survive many more, a narrative of the most involuntary magnificence, variegated, savage, raw, amusing, amorous and disreputable, seething with life, hand in glove with death and the Devil, ultimately contrite and thoroughly tired of a world dissipating itself in blood, pillage and lustfulness, but immortal in the wretched splendour of its sins. Europe is once again in the proper spiritual condition for this book today. There is a great, experienced reading community for it. Perhaps the Swedish edition will give the signal for translations into other European tongues.

Thomas Mann, 1944

What raises Grimmelshausen's work and in particular the Courasche above the contemporary picaresque and rogues' tales, is in the first place the completely original, powerful and artistic language it speaks; in the second place the enchanting presence of its heroine. Both together guarantee the truth of its picture of the age and make it credible to us. Grimmelshausen's language is without peer in the tradition; it is absolutely fresh coinage and has retained its unworn brilliance until today — perhaps just because it has had no successors. Listening to it, one can scarcely believe that it is as old, to the very year, as the Palace of Versailles, as old as Racine's 'Bérénice' and the full-bottomed wig.

Hans Magnus Enzensberger, 1962

The special stamp issued by
the German Federal Post Office
in commemoration of Grimmelshausen Year 1976.

Chronological Table

Grimmelshausen ...		... and his time
	1617	In the Treaty of Prague, King Philipp III of Spain (1598—1621) gives up his right to the Bohemian inheritance in favour of Archduke Ferdinand of Styria, Carinthia and Carniola. Without the consent of the estates, Ferdinand becomes King of Bohemia. He infringes against the rights guaranteed to the Protestants by Emperor Rudolf II (1576—1612) in the Royal Charter of 9 July 1609. Following this, Bohemia suffered from grave unrest. The poet Christian Hofmann von Hofmannswaldau born (died 1679). The "Fruitful Society" founded on 24 August at Burg Hornstein in Weimar.
	1618	"Bohemian-Palatinate War" to 1623. On 23 May the disturbances in Bohemia reach a provisional climax in the Defenestration of Prague: the Imperial Governors Martinitz and Slavata thrown out of the windows of Prague Castle. Count Heinrich Matthias von Thurn (1580—1640) marches against Vienna as leader of the rebels; the Austrian, Silesian, Moravian and Hungarian states under Bethlen Gabor (1580 to 1629) join him. The Duke of Savoy's mercenary leader, Count Ernst von Mansfeld (1580—1626), also fights on the side of the rebellious Protestants.
	1619	The Emperor Matthias (from 1612), brother of Archduke Ferdinand, dies. Ferdinand II crowned new German

Emperor in August in Frankfurt. The Bohemian princes refused to recognize this choice and decide to give the crown to 23 year-old Frederick V of the Palatinate, married since 1613 to Elizabeth, daughter of James I of England (1603—1625) as King of Bohemia. Ferdinand then formed an alliance with Duke Maximilian of Bavaria (from 1597), leader of the "Liga" of Catholic Princes founded in 1609, and with Elector John George I of Saxony (from 1611) and Philipp III of Spain.

Philipp von Zesen (died 1689), founder of the Hamburg "Germanistic Fellowship" and fanatical linguistic reformer, is born. Birth of Cyrano de Bergerac. Publication of a German edition of Thomas Garzoni's "Piazza Universale".

	1620	
The heroine of the novel "Trutz Simplex", Libuschka, known as Courasche, born 1607, escapes from Prachatitz and is present at the Battle of the White Mountain.		Lusatia and Silesia defeated by Saxony, Spanish troops march into the Palatinate under Ambrosius Marquis Spinola (1571—1630). November sees the Battle of the White Mountain near Prague, in which the army of the Liga under Tilly (1559—1632) roundly defeats the Bohemians. The "Winter King" Frederick V of the Palatinate flees to Holland. Aegidius Albertinus (born 1560), translator of Guevara etc., dies. Henry Neville (died 1694) is born. Andreas Perez' "Picara Justina" published in German under the title "Die Landstörtzerin Justina Dietzin".

	1621	
Grimmelshausen, baptized Johann Christoph, born in March in the little Imperial town of Gelnhausen north-east of Frankfurt am Main.		The "Prague Judgment" leads to the execution of 27 instigators of the Bohemian rising, about half of the aristocratic estates are confiscated, 150,000 emigrants leave the country; re-Catholicization and re-Germanization by force. Death in Rome of Pope Paul V, who supported Ferdinand. Succeeded by Gregory XV (1621 to 1623). Philipp III dies in Spain. Philipp IV becomes new King (1621 to 1655). In the Netherlands the War of Liberation, interrupted in 1609, flares up again. Gustav II Adolf, King of Sweden (1623—1632), conquers Livonia.

Death of John Barclay (born 1582). His novel "Argenis" published in the year of his death.

The heroine of the "Abenteuerliche Simplicissimus" born 12 June in Spessart. Courasche's second husband, a captain, falls in the battle of Wiesloch south of Heidelberg. Springinsfeld, hero of the novel of the same name, takes part in this battle at the age of 17 as a drummer boy. Both characters in the novel are involved in the Battle of Höchst.	1622

1622 Tilly storms Heidelberg; the valuable Bibliotheca Palatina collected there since 1560 is given by Maximilian of Bavaria to the Pope as compensation for his expenses in the war. Mansfeld defeats Tilly in the Battle of Wiesloch on 27 April; on 6 May Tilly defeats George Frederick, Margrave of Baden-Durlach (from 1604—1622) at Wimpfen and Prince Christian of Brunswick-Wolfenbüttel (1599—1626) in the Battle of Höchst on 10 June. Molière born. Publication of Charles Sorel's picaresque novel "Francion".

1623 "Lower Saxon-Danish War" to 1629. In Lower Saxony, Protestant Princes and estates form an alliance with Christian IV, King of Denmark and Duke of Holstein (1588—1648). Tilly achieves another victory over Christian of Brunswick at Halberstadt. Electoral dignity bestowed on Maximilian of Bavaria. Death of Pope Gregory XV, succeeded by Urban VIII (1623—1644). Birth of Blaise Pascal. Publication of Campanella's "Civitas solis".

1624 Cardinal Richelieu (1585—1642) becomes Minister to King Louis XIII (1620—1643). Albrecht of Wallenstein (1583—1634) becomes Duke of Friedland. Death of the Silesian cobbler and mystic Jacob Boehme (born 1575). Birth of Johann Scheffler, called Angelus Silesius (died 1677). Publication in Strasbourg of Martin Opitz' poetry collection "Teutsche Poemata" and in Breslau of his epoch-making "Buch von der Deutschen Poeterey".

1625 Wallenstein sets up his own army for Emperor Ferdinand and takes over supreme command of the Imperial troops. Death in England of James I, succeeded by Charles I (1625—1649). Death of Giambattista Marino (born 1569) and John Webster. Publication of Hugo Grotius' "De iure belli ac pacis".

Courasche fights valiantly in the Battle of Lutter am Barenberg, Springinsfeld helps in the same battle to "conquer the King himself".	1626	On 25 April Wallenstein defeats Count Ernst von Mansfeld at the Dessau Bridge, and Tilly defeats Christian IV on 27 August at Lutter am Barenberg. Sigmund von Birken (died 1681), co-founder and later leader of the "Pegnitz Flower Order", is born. Death of Francis Bacon, Lord Verulam (born 1561).
After the death of his father Grimmelshausen's mother re-marries and leaves Gelnhausen. The boy remains in the care of his grandfather Melchior Christoph and attends the Lutheran Latin school.	1627	Tilly and Wallenstein, pressing on to Jutland, occupy the whole of north Germany. Wallenstein becomes Duke of Sagan. Death of Luis de Gongora y Argote (born 1561). Publication of "Los Sueños" by F. de Quevedo Villegas.
	1629	At the height of his power, Ferdinand II issues the Edict of Restitution, requiring the return of all territories which had become Protestant since 1552. The Emperor and King Christian IV conclude the Peace of Lübeck, in which the Danish King — on receipt of his possessions — pledges himself to non-intervention.
Courasche involved in the siege of Mantua in the company of Springinsfeld, with whom she runs her sutler's business.	1630	"Swedish War" to 1635. Wallenstein dismissed at the Electoral Diet of Regensburg, Tilly appointed Supreme Commander of the Imperial Army. On 4 July King Gustav II Adolf of Sweden lands at Usedom. Death of Johannes Kepler (born 1571). First dramatization of the Don Juan material, Tirso de Molina's "El burlador de Sevilla".
Springinsfeld serves in the Swedish army.	1631	In the Treaty of Bärwalde Richelieu guarantees the Swedish King French support. On 20 May Tilly and Pappenheim conquer Magdeburg, which is burned to the ground. On 17 September Tilly defeated in the Battle of Breitenfeld by Gustav Adolf, who had allied himself with John George I of Saxony. The Saxons occupy Bohemia, the Swedes advance as far as Mainz. Birth of John Dryden (died 1700).

In February, after the at-tack on his parents' farm, Simplicius comes to the hermit.

1632   Tilly fatally wounded in Gustav Adolf's victorious battle at Rain am Lech. Wallenstein, re-appointed Imperial Supreme Commander with special powers, reconquers Bohemia. Gustav Adolf occupies Munich with his Commander Johann Banér (1596 to 1641). Wallenstein totally defeated in the Battle of Lützen on 16 November. Gustav Adolf falls in this battle. Duke Bernhard of Saxony-Weimar (1604—1639, Swedish General from 1631) commands the conclusion of the battle. Gustav Adolf's daughter Christine (1626—1683, abdicated 1654 and converted) succeeds to the Swedish throne. Axel Count of Oxenstierna (1583—1654) becomes Head of Government.
Birth of poet and editor of the first German dictionary, Caspar Stieler (died 1707). Birth of John Locke (died 1704) and Baruch Spinoza (died 1677).

1633   Oxenstierna concludes the Heilbronn Treaty with the evangelical Princes "for German liberty and to the satisfaction of Sweden". Duke Bernhard conquers Regensburg. Wallenstein conducts autonomous peace negotiations with Sweden and Saxony and is dismissed and proscribed.
Birth of the novelist Duke Anton Ulrich of Brunswick (died 1714) and of the author of spiritual poems, Catharina Regina of Greiffenberg (died 1694). Publication in Frankfurt of the first volume of "Theatrum Europaeum" and in Breslau of Martin Opitz' "Trostgedichte in Widerwärtigkeit des Kriegs". Foundation in Strasbourg of the "Honest Company of the Fir Trees".

Grimmelshausen takes refuge in the fortress of Hanau near Frankfurt from the Croats plundering Gelnhausen. *Simplicius comes to Hanau after the death of the hermit and becomes page to Governor Ramsay. Springinsfeld takes part in the Battle of Nördlingen, Courasche once again loses a husband in this encounter.*

1634

Wallenstein murdered by the Irish Colonel Butler on 25 February at Eger. On 6 September the united Imperial, Spanish and Bavarian troops totally defeat the Swedes under Duke Bernhard at Nördlingen, under Matthias Gallas, Count of Campo and Duke of Lucera (1584—1647) and Johann von Werth (1592/1602—1652). In September J. v. Ramsay appointed by Bernhard as Governor of Hanau.

In Hanau Grimmelshausen is actually taken by Croats, brought to Hersfeld seminary and probably transferred from Hessen to Cassel. *The same thing happens to the fictional hero Simplicius.*

1635

"Franco-Swedish War" to 1648.
Emperor Ferdinand II concludes the Peace of Prague, with which almost all the Protestant estates associated themselves, with John George of Saxony — waiving the restitution edict. France declares war on Spain and concludes a Treaty of Subsidies with Bernhard of Weimar.
Lope Félix de Vega Carpio (born 1562) dies. Birth of dramatist Daniel Caspar von Lohenstein (died 1683) and Philipp Jakob Spener (died 1705), the subsequent Pietist. First performances of Tirso de Molina's comedy "Don Gil de las calzas verdes" and Calderon de la Barca's "La vida es sueño".

Grimmelshausen witnesses the siege and defeat of Magdeburg as a baggage boy and probably takes part in the Battle of Wittstock. *Simplicius was also at Magdeburg and Wittstock.*

1636

On 4 October the Swedes under Johann Banér defeat the Imperial Saxon Army at Wittstock. The Swedish Commander Linnard Count zu Ortola Torstensson (1603—1651) interrupts an advance on Vienna.
Birth of Daniel Speer (died 1707), author of the "Ungarischen oder Dacianischen Simplicissimus" (1683), and of composer Dietrich Buxtehude (died 1707). First performance of Corneille's tragi-comedy "Le Cid".

Grimmelshausen is settled on Westphalian soil in the regiment of Life Dragoons of Count Hans von Götz, stationed in Soest. *Simplicius guards the Paradise Monastery in the winter of 1636/37 and becomes the "Huntsman of Soest".*

1637

Death of Emperor Ferdinand II. Succeeded by the peace-loving Ferdinand III (to 1657).
Death of Ben Jonson (born 1573). Publication of Descartes' "Discours de la méthode".

Grimmelshausen marches with the Regiment of Count Götz to the relief of the Fortress of Breisach on the Upper Rhine. *Simplex travels to Cologne and Paris, becomes a quack and finally a musketeer in Philippsburg; forced to dig trenches outside Breisach.*

1638   France declares war on Emperor Ferdinand III. Duke Bernhard conquers the fortress of Breisach in December after a long siege.
Martin Opitz translates Sidney's "Arcadia".

Grimmelshausen becomes a musketeer in the Regiment of the Imperial Colonel Hans Reinhard von Schauenburg.

1639   Death of Duke Bernhard of Weimar. Martin Opitz, the most important German poet and literary theoretician of his day (born 1597), dies of the plague. Death of Tomaso Campanella (born 1568). Birth of Jean Baptiste Racine (died 1699).

Springinsfeld enters the Hessian service.

1640   Frederick William, the Great Elector, becomes Elector (to 1688) following the death of George William of Brandenburg.
Death of the poet Paul Fleming (born 1609).

1641   Johann Banér advances on Regensburg and dies. Brandenburg and Sweden conclude an armistice.
Publication of Georg Philipp Harsdörffer's "Frauenzimmer Gesprechspiele" (Part 1) and Christian Gueintz' "Deutscher Sprachlehre Entwurf".

1642   Torstensson defeats the Imperial Troops at the Battle of Breitenfeld. Death of Cardinal Richelieu. His policies continued by Cardinal Jules Mazarin (1602—1661).
Death of Galileo Galilei (born 1564). Birth of poet and novelist Christian Weise (died 1708) and Isaac Newton (died 1727). Part 1 of the "Visiones de Don Quevedo. Wunderbarliche vnd Warhafftige Geschichte Philanders von Sittewald" by Johann Michael Moscherosch published in Strasbourg. Sittewald is an anagram of Willstädt, Moscherosch's birthplace.

	1643	Death of Louis XIII, succeeded by Louis XIV, the "Sun King" until 1715. The Bavarian General Franz Freiherr von Mercy (1590—1645) with Johann von Werth halts the advance of the French at Tuttlingen. First performance of Corneille's "Polyeucte". Publication in Cologne of the German translation of Antonio de Guevara's "Mühseeligkeit dess Hoffs vnd glückseeligkeit dess Landlebens".
	1644	Peace negotiations with the French begin in Münster, with the Swedes in Osnabrück. Turenne and Condé fighting in south Germany. Death of Pope Urban VIII, succeeded by Innocent X (1644—1655). Birth of Ulrich Megerle (died 1709), who became a powerful preacher as a Capuchin under the name of Abraham a Sancta Clara.
Grimmelshausen clerk at regimental headquarters in Offenburg. *Simplicius wounded in the Battle of Jankau, Springinsfeld fighting in Swabia.*	1645	Turenne victorious at Alerheim, Torstensson at Jankau. Hans, Count of Götz, falls in this battle. Death of Hugo Grotius — actually de Groot, Dutch scholar and statesman (born 1583) and Francisco de Quevedo Villegas (born 1580). Publication of Christian Gueintz' "Die Deutsche Rechtschreibung" and Philipp von Zesen's novel "Adriatische Rosemund".
Springinsfeld stays in Augsburg.	1646	Birth of Gottfried Wilhelm von Leibniz (died 1716), the last polyhistorian, founder of the Royal Prussian Academy of Sciences (1700).
	1647	Georg Philipp Harsdörffer's "Poetischer Trichter" (Part 1) published in Nuremberg and Adam Olearius' "Offt begehrte Beschreibung der Newen Orientalischen Reise" in Schleswig. Publication of Johann Rist's "Das Friede wünschende Teutschland" and Eberhard von Wassenberg's "Erneuerter Teutscher Florus". Baltasar Gracián publishes his "Oráculo manual".

Grimmelshausen becomes Staff Secretary in the Regiment of Freiherr von Elter, marches with him to Bavaria and stays in Wasserburg am Inn. *Springinsfeld also comes to Wasserburg.*	1648	The Peace of Westphalia concluded in October in Münster and Osnabrück. Death of Torso de Molina (born 1571).
Grimmelshausen returns to Offenburg and marries Catharina Henninger in August. He then becomes Steward to his erstwhile Regimental Commander Hans Reinhard von Schauenburg in Gaisbach near Oberkirch in the Black Forest.	1649	A great peace banquet takes place in Nuremberg. Publication of Mlle de Scudéry's novel "Cyrus".
	1650	The Swedish troops leave Germany. Ludwig Prince of Anhalt-Köthen (born 1579), co-founder and first supreme head of the "Fruitful Society" and their grammarian Christian Gueintz (born 1592) die. Death of René Descartes (born 1596).
	1651	Death of Elector Maximilian I of Bavaria. His successor Ferdinand Maria (1651—1679) expands Munich into the Residence city. Birth of Quirinus Kuhlmann (burned 1689 in Moscow), an eccentric mystic. Publication of Thomas Hobbes' (1588 to 1679) "Leviathan".
Grimmelshausen buys the land known as the "Spitalbühne" in Gaisbach, on which he builds two houses.	1653	Overthrow in France of the Fronde, the opposition to the absolute monarchy, set up since 1648 by Parliament, the Parisian population and the nobility (Cardinal de Retz [1613 to 1679] and Prince de Condé).
	1654	Friedrich von Logau's "Salomons von Golaw Deutscher Sinn-Getichte Drey Tausend" published in Breslau.
	1655	Death of the aphorist Friedrich von Logau (born 1604). Death of Cyrano de Bergerac. Birth of Johann Beer (died 1700).
	1656	Publication of Mlle de Scudéry's "Clélie".

Grimmelshausen opens the inn "Zum Silbernen Stern" in Gaisbach, which he runs for two years.	1657	Angelus Silesius publishes "Geistreiche Sinn- und Schlussreime" in Vienna.
	1658	Initiated by the Elector and Archbishop of Mainz Johann Philipp von Schönborn (1605—1673), the first Rhine Alliance originates in Frankfurt am Main. This alliance of South-West German Princes, of Brunswick-Lüneburg and Bremen-Verden with France, directed against Austria, is supposed to maintain the provisions of the Peace of Westphalia. Leopold I, son of Ferdinand III, becomes German Emperor.
	1659	With the Peace of the Pyrenees of 7 November, the belligerent confrontations between Spain and France which had been going on since 1635 come to an end. France receives the Catalan county of Roussillon and with it a fixed Pyrenean frontier. Death of the East Prussian poet Simon Dach (born 1605).
Grimmelshausen leaves the service of the Schauenburgs and stays in Gaisbach.	1660	Caspar Stieler's volume of poems "Die geharnischte Venus" appears in Hamburg. Birth of Daniel Defoe (died 1731).
	1661	Death of Cardinal Mazarin. Louis XIV takes over the business of government himself. Building begun on Palace of Versailles, completed in 1684.
Grimmelshausen becomes castellan at the Ullenburg which belongs to the fashionable Strasbourg doctor Johann Küffer. From there he probably makes contacts with Strasbourg.	1662	Death of Blaise Pascal. German translation of Charles Sorel's "Francion". First performance of Molière's "École des femmes".

	1663	Assembly in Regensburg of the Reichstag in permanent session, with representatives of 8 Electors, 165 Princes and 61 Imperial cities participating. Publication in Brunswick of the epoch-making grammatical work "Ausführliche Arbeit von der Teutschen Haubt Sprache" of Justus Georg Schottelius. Andreas Gryphius publishes his "Freuden- und Trauer-Spiele auch Oden und Sonnette" in Breslau. The heroic epic "Hudibras" of Samuel Butler (1612—1680) begins to appear.
Springinsfeld marches against the Turks.	1664	On 1 August the Imperial Commander Raimund Montecuccoli (1609 to 1680) defeats the Turks at St. Gotthard an der Raab, thus ending the first Turkish War which had lasted since 1662. Death of Andreas Gryphius (born 1616), the most important German baroque dramatist.
Grimmelshausen leaves service with Dr. Küffer and settles down in his inn "Zum Silbernen Stern" in Gaisbach.	1665	Birth of Christian Reuter (died 1712), author of the picaresque novel "Schelmuffsky Curiose und Sehr gefährliche Reissebeschreibung zu Wasser und Land" (1696). Publication of Molière's "Don Juan" and La Rochefoucauld's "Maximes".
Part 1 of the "Satyrischer Pilgram" published by G. H. Frommann in Leipzig. In Nuremberg Wolf Eberhard Felssecker publishes the novel "Keuscher Joseph".	1666	The disputed succession of Jülich-Cleves, over which the Protestant electorate of Brandenburg and the Catholic palatinate Neuburg had been opposed since 1609, settled in favour of Brandenburg. Publication of Molière's "Misanthrope" and Boileau's "Satires".
In June Grimmelshausen becomes Strasbourg episcopal Mayor in Renchen in the Black Forest. Frommann publishes Part 2 of the "Satyrischer Pilgram" with Part 1.	1667	The founder of the short-lived "Order of the Elbe Swan", the parson, poet and dramatist from Wedel, Johann Rist (born 1607) dies. Birth of Jonathan Swift (died 1745). Publication of Racine's "Andromaque", John Milton's epic "Paradise Lost".

"Der abenteuerliche Simplicissimus Teutsch" published by W. E. Felssecker in Nuremberg.

1668

End of the so-called War of Devolution between Louis XIV and Spain, or the triple alliance of Holland, England and Sweden in the Peace of Aix-la-Chapelle. In this two-year war Turenne had occupied part of Flanders and the Hennegau and Condé had occupied the Franche-Comté, which France ultimately returned to Spain.
Publication of Henry Neville's "Isle of Pines" in German under the title "Die neu entdeckte Insul Pines".
Publication of "Fables" by La Fontaine.

Felssecker brings out a second edition of the "Simplicissimus" and a "Continuatio". Also the "1. Europäischen Wundergeschichtenkalender". The "Schulmeister Simplicissimus" pirated by Georg Müller in Frankfurt. *Springinsfeld loses a leg in the Battle of Kandia, in which he takes part as a sergeant, and returns to Germany.*

1669

On 27 September the Turks conquer Crete, which had previously been Venetian, with its capital of Kandia.
Death of the satirist Johann Michael Moscherosch (born 1601). Publication in Nuremberg of the first of the five parts of the novel "Die Durchleuchtige Syrerin Aramena" by Duke Anton Ulrich of Brunswick. First performance of Molière's "Tartuffe".

Felssecker publishes the courtly novel "Dietwalds und Amelinden anmuthige Lieb- und Leids-Beschreibung", "Simplicianischer Dreykopffiger Ratio Status", the two simplicianic novels "Trutz Simplex: Oder Ausführliche und wunderseltzame Lebensbeschreibung der Ertzbetrügerin und Landstörtzerin Courasche" and "Der seltzame Springinsfeld" as well as the two shorter works "Der erste Beernhäuter" and "Simplicissimi wunderliche Gauckel-Tasche".

1670

France occupies Lorraine. By an alliance with England — followed by one with Sweden in 1672 — Louis XIV bursts the Triple Alliance and concludes aid treaties with the Bishops of Cologne and Münster in preparation for the second war of conquest against Holland.
Publication in Amsterdam of Philipp von Zesen's Joseph novel "Assenat" and Blaise Pascal's "Pensées" in the Port-Royal edition. Birth of Bernard de Mandeville (died 1733).

Felssecker publishes the "Ewig-währenden Calender" and the "Barock-Simplicissimus".

1671

Death of the poet and travel writer Adam Olearius (born ca. 1599). Birth of Anthony Ashley Cooper Earl of Shaftesbury (died 1713). Publication in Jena of the "Himmlische Liebes-Küsse" of the enthusiast Quirinus Kuhlmann and in Paris of the "Exposition de la Foi catholique" by Bossuet, later Bishop of Meaux.

In Nuremberg Felssecker publishes "Des Abenteuerlichen Simplicii verkehrte Welt". In the course of this year Grimmelshausen probably broke with his Nuremberg publisher. "Rathstübel Plutonis Oder Kunst Reich zu werden" still published by Felssecker, but Part 1 of "Das wunderbarliche Vogel-Nest" and "Des Durchleuchtigen Printzen Proximi, und Seiner ohnvergleichlichen Lympidae Liebs-Geschicht-Erzehlung" are now published by the Strasbourg publisher Georg Andreas Dollhopf. *The second possessor of the magic bird's nest which confers invisibility is severely wounded in July, probably in the fighting when the French crossed the Yssel in Holland.*

1672

France begins the second war of conquest against Holland which lasts until 1678 (Peace of Nymwegen). At first the Netherlands' only ally is Brandenburg-Prussia, later the Emperor and Spain enter the alliance. Turenne, Condé and the King himself conquer the south of Holland at the head of 100,000 men, including 20,000 Germans. The leader of the Dutch "Regent's Party", Jan de Witt, overthrown. William III of Orange (1650—1702) installed at the head of the Republic in 1674.

Death of the composer Heinrich Schütz (born 1585). First performances of Molière's "Femmes savantes" and John Dryden's comedy "Marriage à la Mode".

The linguistic satire "Dess Weltberuffenen Simplicissimi Pralerey und Gepräng mit seinem Teutschen Michel", published by Felssecker or Dollhopf and the "Bart-Krieg oder Des ohnrecht genanten Roht-Barts Widerbellung gegen den weltberuffenen Schwartzbart des Simplicissimi" and "Simplicissimi Galgen-Männlin" published in Strasbourg.

1673

The district of Oberkirch in which Renchen is situated becomes the focus of the south German theatre of war. At the beginning of January the Imperial Cavalry Regiment of Schneidau under Colonel Gundula (Goudela) takes up quarters in Renchen and Sassbach with six companies of 1,000 men each.

Death of Molière.

	1674	Turenne fights Montecucculi on the Upper Rhine. In June troops of Duke Karl of Lorraine (1604—1675), on the Emperor's side, occupy the district of Oberkirch. In December the troops of Elector Karl Ludwig of the Palatinate (1617—1680) and of the Count of Solms make their winter quarters in this area, followed by violence and looting.
Dollhopf in Strasbourg publishes "Dess Wunderbarlichen Vogelnessts zweiter Teil", Grimmelshausen's last work.	1675	In the Battle of Fehrbellin the Great Elector Frederick William defeats the Swedes under Charles X (1660—1697) on 18 June. He had attacked Brandenburg on behalf of France. Before and after the Battle of Sassbach Renchen is the immediate theatre of war. On 23 and 24 July Turenne occupies the place of which Grimmelshausen is Mayor. On 27 July Turenne meets Montecucculi at Achern and falls in the battle at the Renchen Loch near Sassbach.
Grimmelshausen dies in Renchen on 17 August after once again becoming involved in the war.	1676	Deaths of the most important German grammarian of his time, Justus Georg Schottelius (born 1612) and of the ecclesiastical song writer Paul Gerhardt (born 1607). Publication in Halle of Johann Beer's novel "Der Simplicianische Welt-Kucker".

Grimmelshausen's seal to a document,
Gaisbach 2 July 1652.

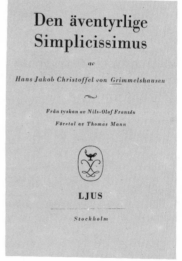

Title pages of translations of Grimmelshausen's works: Om den förträfflige Kyske Josephs ... Lefwernes-Beskriftning, Stockholm 1690; Hôrô no Onna Petenshi Courasche, Tokyo 1967; Simplicissimus, Leningrad 1967; Den äventyrlige Simplicissimus, Stockholm 1944.

Bibliography

1. The most important German Editions

Gesammelte Werke in Einzelausgaben. In co-operation with Wolfgang Bender and Franz Günter Sieveke. Ed. Rolf Tarot, Tübingen: Niemeyer 1967 ff.
Der abentheuerliche Simplicissimus Teutsch und Continuatio des abentheuerlichen Simplicissimi. Ed. Rolf Tarot, 1967.
Des Durchleuchtigen Printzen Proximi und Seiner ohnvergleichlichen Lympidae Liebs-Geschicht-Erzehlung. Ed. Franz Günter Sieveke, 1967.
Dietwalds und Amelindes anmuthige Lieb- und Leids-Beschreibung. Ed. Rolf Tarot, 1967.
Lebensbeschreibung der Ertzbetrügerin und Landstörtzerin Courasche. Ed. Wolfgang Bender, 1967.
Simplicianischer Zweyköpffiger Ratio Status. Ed. Rolf Tarot, 1968.
Des Vortrefflich Keuschen Josephs in Egypten Lebensbeschreibung samt des Musai Lebens-Lauff. Ed. Wolfgang Bender, 1968.
Der seltzame Springinsfeld. Ed. Franz Günter Sieveke, 1969.
Satyrischer Pilgram. Ed. Wolfgang Bender, 1970.
Das wunderbarliche Vogelnest. Ed. Rolf Tarot, 1970.
Kleinere Schriften. Ed. Rolf Tarot, 1973.
Die verkehrte Welt. Ed. Franz Günter Sieveke, 1973.

Der abenteuerliche Simplicissimus. Ed. Alfred Kelletat, München: Winkler 1956, ²1967.
Simplicianische Schriften. Ed. Alfred Kelletat, München: Winkler 1958.

2. Grimmelshausen in Translation

Belgium
De avontuurlijke Simplicissimus. Werd in den nieuwe bewerking van Frans Lichtenberger. Uit het Duitsch vertaald door L. Raeymaekers, Leuven: Davidsfonds 1939 (= Volksreeks 287).
De avontuurlijke Simplicissimus. Bewerkt voor de jeugd door Frans Willems (i. e. Franciscus Antonius Brunklaus). Geill. door Bob Man, Retie (Belgie): Kempische boekhandel 1957.

Czechoslovakia
Dobrodružný Simplicius Simplicissimus. Kronika třicetileté války. Z něm. orig. přel. Jaroslav Zaorálek. Dobové obrázky podle současných

rytin překreslil Vojtěch Kubasta, vad. 2, Praha: Vyšehrad 1951. (= Živý Odkaz světa. Sv. 17).

Vojna. Trilogie z třicetileté války. Z něm. predlohy přel., úvodní studii a poznámky napsal Pavel Eisner. Na přilohách 14 rytin Jacquesa Callota, Praha: Melantrich 1953.

(Selection from the "Simplicianischen Schriften")

Simplicius Simplicissimus. Kronika třicetileté války. Z něm. orig. »Der abenteuerliche Simplicissimus«, přel. Jaroslav Zaorálek. Překlad upravil a doplnil, vysvetl. i doslov naps. Bohumil Novák. Ill. Karel Toman, vyd. 2, Praha: Naše vojsko 1959.

Dobrodružný Simplicius Simplicissimus. Čiže životopis podivného vaganta memon Melchior Sternfels ... na svetlo vyd. German Schleifheim von Sulsfort. Přel. z něm. orig. a poznámky spracoval Ján Belnay. Doslov napísal Július Pašteka, Verše prebásnil Viliam Turčány. Drevorezy od Ernesta Zmetáka, Bratislava: Slovenské Vyd. kránej literatúri 1964.

Poběhlice Kuráž. Divous Skočdopole. Z něm. orig. přel. a poznámky napsal Rio Preisner. Predmluvu napsala Brigitta Wolfová. Verše přel. Josef Suchý, Praha: Odeon 1968. (= Světová Četba Sv. 399).

Denmark

Den eventyrlige Simplicissimus. Overs. fra tysk af Mogens Boisen, København: Hans Reitzel 1963, 3 vol. (= Reitzels Serie 113—115).

Esperanto

La aventuoj de Simplicius Simplicissimus, in: Germana Esperanto, Revuo Juni 1974—Dezember 1974, wieder in: Europa Esperanto, Revuo März 1975—Dezember 1975. (Teilübersetzung der Esperanto-Gruppe Oberkirch [Hermann Heiss, Margot Hektor, F. W. Poppeck]).

Finland

Seikkailukas Simplicissimus. Rehevän vallaton kuvaus kolmikymmen-vuotisen sodan ajoilta. Suom. Werner Anttila. Edellinen ja jälkimmäinen osa, Hämeenlinna: Arvi A. Karisto 1950 (= Kariston klassillinen kirjasto 61—62).

France

La vagabonde Courage. Traduit de l'allemand par M. Colleville; précédé d'une notice historique et critique, Paris: La Renaissance du Livre 1922 (Collection de littérature ancienne française et étrangère).

Les aventures de Simplicius Simplicissimus. Ouvrage traduit de l'allemand par M. Colleville ... et précédé d'une préface ..., 2 Bde, Paris: La Renaissance du Livre 1926 (Collection de littérature ancienne française et étrangère).

Les aventures de Simplicius Simplicissimus. Traduction et introduction de Maurice Colleville, Paris: Sullivier 1951 (= Fiction et Vérité 103).

Les aventures de Simplicius Simplicissimus. Traduction, introduction et notes par Maurice Colleville ..., Paris: Éditions Montaigne 1963 (Collection bilingue des classiques étrangers), 2 Bde.

La vagabonde Courage. Traduit de l'allemand et présenté par Maurice Colleville, Paris: Union générale d'éditions 1963 (= Le Monde en 10—18, 139).

Great Britain
The adventurous Simplicissimus, being the description of the life of a strange vagabond named Melchior Sternfels von Fuchshaim. Written in German by Hans Jacob Christoph von Grimmelshausen, and now for the first time done into English. Translated by A. T. S. Goodrick, London: W. Heinemann 1912.

Simplicissimus the vagabond, that is the life of a strange adventurer named Melchior Sternfels von Fuchshaim: namely where and in what manner he came into this world, what he saw, learned, experienced, and endures therein; also why he again left it of his own free will. Exceedingly droll and very advantageous to read. Given forth by German Schleifheim von Sulsfort in the year MDLXIX, translated by A. T. S. Goodrick ... with an introduction by William Rose ..., London: G. Routledge & son, New York: E. P. Dutton & Co. 1924.

The adventures of a simpleton (Simplicius Simplicissimus) newly translated from the German by Walter Wallich, London: The New English Library 1962 (= Four square classics 1003).

Simplicius Simplicissimus. Translated from the original German edition of 1669 by Hellmuth Weissenborn and Lesley Macdonald, with engravings by Hellmuth Weissenborn. Verses translated by David Rodger, London: John Calder 1964.

Mother Courage. Translated by Walter Wallich, liftground drawings by Fritz Wegner, London: Folio Society 1965.

Hungary
A kalandos Simplicissimus. Übers. v. Julius Hay. Ill. v. Hincz Gyula. Anm. u. Nachw. v. Lajos Pók, 2 Bde, Budapest: Magyar Helikon Kiadó 1964.

Italy
L'avventuroso Simplicissimus. Prima edizione italiana. Traduzione di Angelo Treves, Milano: Monanni 1928.

L'avventuroso Simplicissimus. Traduzione e introduzione di Camilla Conigliani, Torino: U.T.E.T. 1945 (I grandi scrittori stranieri).

L'avventuroso Simplicissimus. Traduzione dal tedesco di Ugo Dèttore e Bianca Ugo. Introduzione di Ugo Dèttore. Illustrazioni di Fulvio Bianconi, Milano: Bianchi-Giovini 1945 (Aretusa. Narrativa classica italiana e straniera).

L'avventuroso Simplicissimus. Edizione integrale. Traduzione di Ugo Dèttore e Bianca Ugo. Introduzione di Ugo Dèttore, 2 Bde. Milano-Verona: A. Mondadori 1954, ²1956 (Biblioteca Moderna Mondadori, CCCLXXIV—CCCLXXV. Nuova Serie).

L'avventuroso Simplicissimus. A cura di Camilla Conigliani, Torino: U.T.E.T. 1958.

L'avventuroso Simplicissimus. Traduzione di Ugo Dèttore e Bianca Ugo, Milano: Club degli editori 1970.

Japan

Ahô monogatari. Kiyonobu Kamimura yaku, in: Sekai Bungaku Zenshû, Tôkyô: Kawade shobô 1951, S. 221—282.
Ahô monogatari. Ichie Mochizuki yaku, Tôkyô: Iwanami shoten 1953 (= Iwanami Bunko 5026—28).
Hôrô no me-peten-shi Kurâshe. Nakada Miki yaku, Tôkyô: Gendai Shichô-sha 1967 (= Koten Bunko 15).

Netherlands

De avontuurlijke Simplicissimus. Nieuw verteld door Anton Thiry, Amsterdam: Uitg.-Mij. „Prometheus" 1925.

Poland

Przygody Simplicissimusa. Tłum. Anna Maria Linke. Wiersze tłum. J. Dackiewicz, Warszawa: Panstw. Instytut Wydawn 1958 (Biblioteka Arcydzieł. Najsławniejsze Powieści Świata).

Rumania

Aventurosul Simplicius Simplicissimus I, II. În românește de Virgil Tempeanu. Pref. de Livia Stefanescu, Bukareşti: Ed. pentru literatura 1967 (= Biblioteca pentru toti 401—02).

Sweden

Om den förträfflige Kyske Josephs, Jacobs Kiäreste Son i Egypten, Upbyglig, utförlig, myckel tilökt, och berömlige Lefwernes-Beskrifning ... Af Samuel Greifn-Sohn von Hirschfeldt Och förswänskat, (Stockholm) 1690.
Den Durchleuchtigsta Prinsens Proximi och Hans Oförlikneliga Lympide Kärleks-Händelsers Beskrifning ... Utdrag utur H. J. Christoffel von Grimmelshausen, Norrköping: Johan Edman 1763.
Den kyske Joseph Patriarchen Jacobs Son, I Deß Förnedrade och uphögda Tillstånd, Såsom et Mönster för alla af wälsinta Föräldrar födda och upfödda barn ..., Uppsala: Johan Edman 1767.
Den äventyrlige Simplicissimus. Från tyskan av Nils-Olof Franzén. Företal av Thomas Mann, Stockholm: Ljus 1944.
Den äventyrlige Simplex Simplicissimus. Ett soldatöde från 30-åriga kriget ... Pedagogisk bearb.: Holger Nilsson. Illustr. av. Harald Hägg, Stockholm: Liber 1961 (Vår skolas bibliotek).

Union of Soviet Socialist Republics

Cudakovatyj Simplicissimus ili Opisanie žizni odnogo cudaka po imeni Mel'chior Šternfel'd fon Fusgejm. Per. i obrabotka E. G. Guro. Pod red. i s predisl Je. Lanna, Moskwa, Leningrad: Zemlja i fabrika 1925.
Simplitsissimus. Izdanie podgotovil Aleksandr Morozov. Perevod A. A. Morozova i E. G. Morozovoi, Leningrad: Izd-vo Nauka 1967 (Literaturnye Pamjatniki).

United States of America

The adventurous Simplicissimus, being the description of the life of a strange vagabond named Melchior Sternfels von Fuchshaim. Translated by A. T. S. Goodrick, Lincoln: University if Nebraska Press 1962.

94

The adventures of a simpleton (Simplicius Simplicissimus). Newly translated from the German by Walter Wallich, New York: Ungar 1963. (= Ungar Paperbacks 2129).
Satyrischer Pilgram; a critical edition with introduction and notes by Mary Louise Klein, Diss. Austin, University of Texas 1963.
Courage, the adventuress & The false messiah. Translation and introduction by Hans Speier, Princeton N. J.: Princeton University Press 1964.
Simplicius Simplicissimus. A modern translation, with an introduction by George Schulz-Behrend, Indianapolis: Bobbs-Merrill 1965. (= The library of liberal arts. Literature 186).
The Runagate Courage. Spite Simplex or the detailed and wondrously strange life history of the archfraud and runagate Courage ... Translated by Robert L. Hiller and John C. Osborne, Lincoln: Univ. of Nebraska Press 1965.

Yugoslavia
Simplicissimus. Preveo Nikola Popović. Stihove prejevao Branimir Živojinović, Beograd: Prosveta 1953 (Svetski Klasici).
Pustolovni Simplicissimus. Übers. v. Marija Kon, Sarajevo: Svjetlost 1955.
Simplicius Simplicissimus. Prevedel in Spremno besedo napisal Ivan Stopar. Oprema: Jakica Accetto, Ljubljana: Cankarjeva zalozba 1961 (Svetovni Roman).
Simpilicius Simplicissimus. Prevod v. Nikola Popović. Ill. v. Maks Hunciker, Beograd: Jugoslavija 1967 (Biblioteka Sudbine).

3. The most important literature on Grimmelshausen

Scholte, Jan Hendrik: Probleme der Grimmelshausen-Forschung, Groningen 1912.
Bechtold, Arthur: Grimmelshausen und seine Zeit, Heidelberg 1914, ²1919.
Könnecke, Gustav: Quellen und Forschungen zur Lebensgeschichte Grimmelshausens. Published for the Gesellschaft der Bibliophilen by J. H. Scholte, 2 Vols., Leipzig 1926/28.
Koschlig, Manfred: Grimmelshausen und seine Verleger. Untersuchungen über die Chronologie seiner Schriften und den Echtheitscharakter der frühen Ausgaben, Leipzig 1939 (= Palaestra 218). Reprint New York, London 1967.
Scholte, Jan Hendrik: Der Simplicissimus und sein Dichter. Gesammelte Aufsätze, Tübingen 1950.
Streller, Siegfried: Grimmelshausens simplicianische Schriften. Allegorie, Zahl und Wirklichkeitsdarstellung, Berlin 1957 (= Neue Beiträge zur Literaturwissenschaft 7).
Weydt, Günther: Nachahmung und Schöpfung im Barock. Studien um Grimmelshausen, Bern, München 1968.
Weydt, Günther, ed.: Der Simplicissimusdichter und sein Werk, Darmstadt 1969 (= Wege der Forschung CLIII).

Weydt, Günther: Hans Jacob Christoffel von Grimmelshausen, Stuttgart 1971 (= Sammlung Metzler 99).

Battafarano, Italo Michele, Grimmelshausen-Bibliographie 1666—1972. Werk — Forschung — Wirkungsgeschichte. Unter Mitarbeit von Hildegard Eilert, Napoli 1975. (= Quaderni degli annali dell'istituto universitario orientale. Sezione germanica 9)

Berghaus, Peter and Günther Weydt: Simplicius Simplicissimus. Grimmelshausen und seine Zeit (Ausstellungskatalog). Münster 1976.

Koschlig, Manfred: Das Ingenium Grimmelshausens und das Kollektiv. Studien zur Entstehungs- und Wirkungsgeschichte des Werks, München 1976.

4. Foreign-language literature on Grimmelshausen in chronological order

Ernuf: Nemeckij Žilblaz XVII veka, in: Zagraničnyi vestnik 1 (1865).

Antoine, Ferdinand: Étude sur le Simplicissimus de Grimmelshausen. Thèse française, Paris 1882.

Arrighi, Giuseppe: Il Sacchetti tedesco, in: Fanfulla della Domenica 26 (1904) Nr 26, 26. Juni.

Bossert, Adolphe: Le roman de la guerre de trente ans: le »Simplicissimus«, in: A. B., Essais sur la littérature allemande, Paris 1905, Bd 1, S. 4—52.

Bourdeau, Jean: Le Simplicissimus de Grimmelshausen. Un Gil Blas allemand, in: J. B., Poètes et Humanistes de l'Allemagne. La France et les français jugés à l'étranger, Paris 1906, S. 1—15.

Fasola, Carlo: Il capitolo 30, libro 2 del Simplicissimus, in: Rivista di letteratura tedesca 1 (1907) S. 24—33.

Lovell, George Blakemann: Word-Order in the works of Grimmelshausen (1625—1676), as indicated by the position of the Verb, Diss. Yale 1909.

Lovell, George Blakemann: Peculiarities of verb-position in Grimmelshausen, in: The Journal of English and Germanic Philology 11 (1912) S. 205—208.

Kalff, G.: Hooft en Grimmelshausen, in: Tijdschrift voor nederlandse Taal- en Letterkunde 32 (1913) S. 149f.

Waterhouse, Gilbert: The literary relations of England and Germany in the 17th century, Cambridge 1914.

Friedmann-Coduri, Teresita: Un romanzo tedesco del Seicento: Simplicissimus, in: Nuova antologia di scienze, lettere ed arti 259 (1915) S. 434—448.

Lafleur, Paul T.: A seventeenth-century Pan-Germanist, in: Notes and Queries 11 (1915) S. 377.

Sternberg-Montaldi, Federigo: Grimmelshausen ed il suo tempo, in: Rivista d'Italia 1 (1915) S. 3—48.

Sehrt, Edward H.: Grimmelshausen as a dialectologer, in: Modern Language Notes 31 (1916) S. 338—342.

Bottacchiari, Rodolfo: Saggio su l'avventuroso Simplicissimus, Torino 1920.

Gudde, Erwin Gustav: Grimmelshausen's Simplicius Simplicissimus and

Defoe's Robinson Crusoe, in: Philological Quarterly 4 (1925) S. 110—120.

Lann, Je.: Predisl., v. kn.: Cudakovatyi Simplicissimus ili opisanije zizni odnogo cudaka, Moskau, Leningrad 1925.

Bürger, C. P.: Het Eiland der Vruchtbaarheid, in: Het Boek 19 (1930) S. 321—332.

Hayens, Kenneth C.: Grimmelshausen's minor works, in: The Journal of English and Germanic Philology 30 (1931) S. 516—530.

Rose, William: Grimmelshausen and his Simplicissimus, in: W. R., Men, myths and movements in German Literature. A volume of historical and critical papers, London 1931, S. 85—107.

Hayens, Kenneth C.: Grimmelshausen, London 1932 (= St. Andrews University Publications 34).

Fedorov, A.: O »Simplicissimuse«, Grimmelshausen i ego Perevode, in: Rezec 6 (1934).

Scholte, Jan Hendrik: Het probleem van den »Ur-Simplicissimus«, in: De Weegschaal 3 (1936) S. 97—101.

Scholte, Jan Hendrik: Uit het jaar 1672, in: Neophilologus 23 (1938) S. 402—407.

Weigel, Harold Wildie: The romance element in the vocabulary of Grimmelshausen's Simplicissimus, Diss. Pennsylvania State University 1940.

Zieglschmid, A. J. F.: An unpublished »Hausbrief« of Grimmelshausen's Hungarian Anabaptists, in: The Germanic Review 15 (1940) S. 81—97.

Scholte, Jan Hendrik: Een Oestinjevaarder in den Simplicissimus, in: De Gids 106 (1942) Nr 4, S. 69—84.

Scholte, Jan Hendrik: Ons land bij Grimmelshausen, in: Levende Talen (1944).

Scholte, Jan Hendrik: Restauratie van den Simplicissimus, in: Neophilologus 29 (1944) S. 26—33.

Henning, John: Grimmelshausen's british relations, in: Modern Language Review 40 (1945) S. 37—45.

Scholte, Jan Hendrik: Naar aanleiding van een sonnet van Vittoria Colonna, in: Neophilologus 31 (1946) S. 134—138.

Hammer, Carl: Simplicissimus and the literary historians, in: Monatshefte für deutschen Unterricht 40 (1948) S. 457—464.

Gravier, Maurice: La »simplicité« du »Simplicissimus«, in: Études Germaniques 6 (1951) S. 163—168.

Scholte, Jan Hendrik: Robinsonades, in: Neophilologus 35 (1951) S. 129—138.

David, Claude: Un grand roman du 17e siècle, in: Critique 57 (Février 1952) S. 105—119.

Forster, Leonard: »Beau Alman« et le »Beau Escuyer Gruffy«, une source française de Grimmelshausen?, in: Études Germaniques 7 (1952) S. 161—163.

Wichert, Hildegarde E.: Johann Balthasar Schupp and the Baroque satire in Germany, New York 1952.

Bihalji-Merin, Oto: Grimmelshauzen i njegov Simplicissimus, in: Kńiževnost 16 (1953) S. 510—522.

Roy, Claude: Grimmelshausen, in: C. R., Descriptions critiques, Paris 1953, Bd 2, S. 105—109.

Weil, H. H.: The conception of the Adventurer in German Baroque literature, in: German Life and Letters N. S. 6 (1953) S. 285—291.

Landa, Je.: Tridcatiletnjyjy vojna i perspektivy obscestvennogo razvitija Germanii v ocenke nemeckoj narodnoj revoljucionnoj literatury 17 veka. (Osnovnye idejnyje motivy romana Grimmelshausen »Prikljucenija Simplicija Simplicissimusa«), Naucnyje Zapinski Vsesojuznoj vyssej finansovoj skoly, Vyp. 1, C. 2, 1956.

Kaltenbergh, L.: Przygody Simplicissimusa, in: Radio i Świat 12 (1956) Nr 32, S. 4.

Hyde, James Franklin: The religious thought of J. J. C. von Grimmelshausen as expressed in the »Simplicianische Schriften«, Diss. Bloomington Indiana 1960.

Dahl, Etta: Grimmelshausen's »Simplicissimus«, a study of a critical deformation, Diss. Los Angeles, California 1961.

Navarro de Adriaenses, José María: La continuación del »Lazarillo« de Luna y la aventura del Lago Mummel en el »Simplicissimus«, in: Romanistisches Jahrbuch 12 (1961) S. 242—247.

An.: Trials of a simple soldier, in: The Times Literary Supplement, 1962 Nov. 9, S. 849 f.

Morozov, Aleksandr A.: Grimmelshausen i satira XVII veka, v. kn.: Istorija nemeckoj literatury, B 5-ti t., T. 1, Moskau 1962.

Nakata, Kazuhiro: (Über Grimmelshausens »Courasche«), in: Tohoku DBK 6 (1962) (japanisch mit deutscher Zusammenfassung).

Romberg, Bertil: Grimmelshausens »Simplicissimus«, in: B. R., Studies in the narrative technique of the firstperson novel, Stockholm, Göteborg, Uppsala 1962, S. 147—176.

Nakagava, Seizo: Das Problem der 2. Person bei Grimmelshausen, in: Forschungsberichte zur Germanistik 5 (1963) S. 20—33, 83 f. (japanisch mit deutscher Zusammenfassung).

Nakagava, Seizo: Über die Manier des Romans von Grimmelshausen in Bezug auf die Simplicianischen Schriften, in: Die deutsche Literatur, Osaka, 10 (1964) Nr 10, S. 17—37 (japanisch mit deutscher Zusammenfassung).

Villa, Vincenzo Maria: Simplicissimus, in: V. M. V., Postille di letteratura tedesca, Milano-Varese 1964, S. 83—87.

Gilbert, Mary E.: Simplex and the battle of Wittstock, in: German Life and Letters N. S. 18 (1964/65) S. 264—269.

Hiller, Robert L.: The sutler's cart and the lump of gold, in: The Germanic Review 39 (1964) S. 137—144.

Jacobson, John Wesley: Three leaves from the »Gauckel-Tasche« in the works of Grimmelshausen, Diss. Bloomington Indiana 1964.

Frederik, Ezra: The treatment of the jew in works of Buchholz, Grimmelshausen and Happel, Diss. Indiana University 1965.

Magris, Claudio: Le robinsonaden fra la narrativa barocca e il romanzo borghese, in: Arte e Storia. Studi in onore di Leonello Vincenti, Torino 1965, S. 233—284.

Jacobson, John Wesley: The culpable male. Grimmelshausen on women, in: The German Quarterly 39 (1966) S. 149—161.

Nakagawa, K.: Über den »Ewigen« beim »Simplicissimus« — aus der Vorrede des Ewigwährenden Calenders, in: Doitsu Bungaku 36 (1966) S. 50—60 (japanisch mit deutscher Zusammenfassung).

Nakata, Kazuhiro: Über Grimmelshausens Vogelnest, in: Tohoku DBK 10 (1966) S. 19—35 (japanisch mit deutscher Zusammenfassung).

Arnold, Herbert A.: J. J. C. von Grimmelshausen and his translators, in: Monatshefte für deutschen Unterricht, deutsche Sprache und Literatur 59 (1967) S. 351—357.

Jacobson, John Wesley: A defense of Grimmelshausen's Courasche, in: The German Quarterly 41 (1968) S. 42—54.

Heckmann, J. F.: Developments in narrative consciousness in Sorel, Grimmelshausen and Prevost, Diss. Cornell University 1968.

Peters, Judith Griessel: The spanish picaresque novel and the simplician novels of H. J. Chr. v. Grimmelshausen, Diss. University of Colorado 1968.

Rehder, Helmut: Planetenkinder: Some Problems of Character Portrayal in Literature, in: The Graduate Journal VIII, 1, The University of Texas, Austin 1968, S. 69—97.

Rosenfeld, Anatol: Tricentenário de um grande romance, in: O Estado de Sao Paulo Suplemento literário, 1968 Mai 11, S. 1.

Heckmann, John: Emblematic structures in Simplicissimus Teutsch, in: Modern Language Notes 84 (1969) S. 876—890.

Holzinger, Walter: Der abenteuerliche Simplicissimus and Sir Philip Sidney's Arcadia, in: Colloquia Germanica 3 (1969) S. 184—198.

Speier, Hans: Courage, the adventuress, in: H. S., Force and folly. Essays on foreign affairs and the history of ideas, Cambridge, Massachusetts, London 1969, S. 258—278.

Speier, Hans: Grimmelshausen's laugther, in: ibid. S. 279—323.

Speier, Hans: Simplicissimus, the irreverent fool, in: ibid. S. 235—257.

Battafarano, Italo: Il »Simplicissimus« di Grimmelshausen, Diss. Bari 1970.

Nakata, Kazuhiro: Zum Abendlied des Einsiedlers von Grimmelshausen, in: Tohoku DBK 14 (1970) S. 1—18 (japanisch mit deutscher Zusammenfassung).

Nee, James Michael: Myth and the »Bildungsroman«. An archetypal study of Grimmelshausen's Simplizissimus, Diss. University of Michigan 1970.

Gonzáles, Manuel J.: Lo guevariano en el »Simplicius Simplicissimus«, in: Letras de Deusto 1 (1971) Nr 2, S. 83—101.

Knight, K. G.: Grimmelshausen's Simplicissimus — a popular Baroque novel, in: Periods in German literature 2 (1971) S. 1—20.

Battafarano, Italo: Simpliciana utopica. Dall' ascetismo all' idillio pre-rousseauisme, in: Aion 15 (1972) 2, S. 7—36; 3, S. 7—41.

Jones, G. L.: Democritus »versus« Heraclitus: a note on satire and irony in Grimmelshausen and Wieland, in: Trivium 7 (1972) S. 125—128.

Mandel, Siegfried: From the Mummelsee to the moon: refractions of science in seventeenth-century literature, in: Comparative Literature Studies 9 (1972) S. 407—415.

Rehder, Helmut: The last hermitage, in: Traditions and transitions. Studies in honor of Harold Jantz. Ed. by L. E. Kurth, W. E. McClain, H. Homan, München, Baltimore 1972, S. 64—69.

Sheppard, Richard: The narrative structure of Grimmelshausen's »Simplicissimus«, in: Forum for Modern Language Studies 8 (1972) S. 15—26.

Ashcroft, Jeffrey: Ad astra volandum: emblems and imagery in Grimmelshausen's »Simplicissimus«, in: Modern Language Review 68 (1973) S. 843—862.

Sources of illustrations: Frontispiece: A. Paul Weber, Schretstaken near Hamburg; p. 11, 17, 38, 48, 56, 58, 61: Wolfenbüttel, Herzog-August-Bibliothek; p. 20 top, 54: Nuremberg, Germanisches Nationalmuseum; p. 20 bottom: Heimatmuseum Gelnhausen; p. 21, 22: Stadtverwaltung Gelnhausen; p. 25 top, 37: Frankfurt/Main, Stadt- und Universitätsbibliothek; p. 25 bottom, 27, 64: Münster, Landesmuseum; p. 29 left: Offenburg, Ritterhausmuseum; p. 29 right: Freiburg, Private collection; 32, 67: Bonn, Presse- und Informationsamt der Bundesregierung; p. 33, 41: Publisher's archive; p. 34: Stadtbibliothek Mainz; p. 36: dpa; p. 44: Wiesbaden, VMA; p. 50: Münster, Germanistisches Institut der Universität; p. 70, 89: Karlsruhe, Badisches Generallandesarchiv; p. 74: Bonn, Bundespostministerium.

Publisher and author are gratefully indebted to Prof. Dr. P. Berghaus and Prof. Dr. G. Weydt for their generosity in making available the illustrative material assembled for their great commemorative exhibition "Simplicius Simplicissimus. Grimmelshausen and his Time". The author gratefully acknowledges his debt to the publications of G. Weydt, M. Koschlig and I. M. Battafarrano. The quotations were taken from "Die Simplicianischen Bücher", Volume 3, Munich 1925, edited by Kelletat and E. Hegaur (see bibliography).